How Militaries Learn

How Militaries Learn

Human Capital, Military Education, and Battlefield Effectiveness

Nathan W. Toronto

LEXINGTON BOOKS
Lanham • Boulder • New York • London

Published by Lexington Books
An imprint of The Rowman & Littlefield Publishing Group, Inc.
4501 Forbes Boulevard, Suite 200, Lanham, Maryland 20706
www.rowman.com

Unit A, Whitacre Mews, 26-34 Stannary Street, London SE11 4AB

British Library Cataloguing in Publication Information Available

Library of Congress Cataloging-in-Publication Data Available

LCCN 2018947267 | ISBN 9781498571661 (cloth: alk. paper) |
ISBN 9781498571678 (electronic)

∞™ The paper used in this publication meets the minimum requirements of American National Standard for Information Sciences—Permanence of Paper for Printed Library Materials, ANSI/NISO Z39.48-1992.

Printed in the United States of America

To Hannah, Rebecca, Caleb, and Jonathan

May we leave the world better than you found it

Contents

Figures

Tables

Preface

How do militaries learn? The answer to this question has both theoretical and practical implications for security in the world today.

Practically speaking, adopting patterns of military practice that have proven effective elsewhere is a key component of state power, leading most militaries today to regard advising foreign forces as an essential mission. In this major effort, the United States and its allies spent billions of dollars and thousands of hours training and equipping the Iraqi security forces from 2003 on, only to see these forces, in August 2014, fold in the face of modest pressure from ISIS insurgents. Foreign training and equipping efforts in Yemen similarly failed to prepare the armed forces to defend state institutions from rebel Houthi offensives in late 2014 and early 2015. Such stories of marginal military performance, despite significant training efforts, are not unique to Yemen and Iraq. The leading militaries of the world have long provided military advice and training to friends and allies, but these efforts have not always led to military success on the battlefield. Military officers, then, might read this book to understand what explains the success or failure of military advising efforts.

From a theoretical perspective, the problem of acquiring military knowledge goes well beyond military advising to understanding state power, civil-military relations, and the diffusion of ideas among states. States have often sought to build their power by replacing old military institutions, constructed for a bygone security environment, with new ones, more appropriate for a current or future environment. For example, in the late nineteenth century the US military was regarded with disdain in Europe, but now is the gold standard in performance and professionalism. Even though the French adopted the mass army before other European countries, at the end of the eighteenth century, Prussia eventually implemented the mass army system more effectively than France.[1] The Ottoman Empire struggled for over a century to adopt patterns of Western military practice, but only started to experience success in the middle of the twentieth century.[2] Political scientists and historians have often pointed to material factors like technology, interstate competition, political coalitions, and economic growth to explain such power shifts, but few have examined the role of military learning in developing state power.

Understanding how militaries learn can also inform theories of civil-military relations. Scholars of democratization in Latin America have for decades investigated why military forces choose to stay in the barracks and let democracy break out. Likewise, scholars of civil-military relations in other regions examine how or why militaries intervene in politics. Many of the explanations for military intervention or restraint focus on the strength of civil-society institutions, levels of military institutionalization, coalition building, and popular unrest, but rarely do these explanations cross regional or cultural borders. This study makes less of such regional peculiarities, and suggests that, across the world, longer-term historical forces of economic development and human capital growth also moderate conflict between civilian and military authorities.

How militaries learn also has implications for understanding the diffusion of ideas between states. Some scholars of idea diffusion treat all states as equally capable of adopting new military institutions, and treat diffusion as an inexorable gravitational pull toward the most successful methods of organization that the system has to offer at the time.[3] Other scholars relax this black box model of the state to examine the role of elite networks, democratization, or culture in the adoption of new military ideas. Like these scholars, I accept the notion that some states are better at than others at adopting the most successful military institutions available, and offer a theory to explain these differences.

What does it mean to adopt the most successful military institutions available? The "what" in what militaries learn, in other words, is the system of military reflection, a distinct characteristic of mature military organization in the modern era. The system of military reflection is a pattern of addressing military problems with strategic questioning and decentralized execution. It began to manifest itself during the Napoleonic Wars, but some states have proven more successful than others at adopting it. This system of military reflection manifests itself in the existence of (1) a policy role for independent military analysis, (2) intellectual space for critical evaluation of military options, and (3) decentralized execution of military decisions.

In practice, this system of reflection manifests itself in a number of ways in both peacetime and wartime. During wartime, states that have adopted the system of military reflection have a formal system for seeking input on wartime decisions, be it on whether to go to war in the first place or on what the military strategy should be. The military forces of these states also tend to have a formal staff system to evaluate operational decisions, generate orders, and enforce execution of those orders, usually according to a "mission orders" framework. These militaries also manage battlefield information effectively by passing both negative and positive information up the chain in a timely manner, and then by adapting to that information in time to take advantage of opportunities that arise. Militaries that have adopted the system of reflection also reward initiative in promotions and battlefield commendations. A number of scholars have identified these characteristics of how a modern, reflective military fights on the battlefield.[4]

More difficult is how to identify a reflective military before it steps onto the battlefield. The main contention of this study is that one of the best ways to observe the system of reflection in peacetime is to examine military education. As chapter 1 suggests, the institution of staff and war colleges (for officers at the mid- and two-thirds points in their careers, respectively) is an important initial marker of adopting the system of reflection. Also important is how those schools teach what they teach, since teaching by Socratic inquiry, discussion, and critical research is a more faithful marker of the system of reflection than teaching by rote memorization and unquestioned lectures. In addition, militaries that have adopted the system of reflection also have robust training patterns, exercising at regular intervals using realistic battlefield conditions, and clear standards of success and failure. Reflective militaries also tend to publish more periodicals, with dissenting opinions and critical analysis of both historical and contemporary cases. It is often more difficult to identify the system of military reflection during peacetime, but not impossible, as this study demonstrates.

Military institutions reflect these characteristics to varying degrees from country to country. An individual country's military institutions can change over time, and these ideal characteristics can clash with cherished military values of discipline, respect for authority, and loyalty to political leaders. However, history suggests that militaries that come closer to implementing the ideals of the system of reflection are more likely to succeed on the battlefield.[5]

What, then, explains when states will adopt this system of military reflection? The main thesis of this study is that, over the last two hundred years, political will and access to wealth and human capital have been the most important factors in acquiring the system of military reflection.

POLITICAL WILL

Adopting the system of military reflection requires significant political will because it involves a cultural shift: building new schools with new education methods, disempowering old elites, and, potentially, reordering the existing social bargain. For example, in 1826 the Ottoman Sultan slaughtered hundreds of Janissaries with massed cannon fire in order to make political space to develop his Nizam-e Jadid (New Model) army. Not all episodes of adoption have been so dramatic or violent, such as the British Army adopting the Haldane Reforms in the aftermath of battlefield setbacks in the Boer War, which suggests that some political systems may be more adaptive than others. The point is that adopting the system of military reflection takes political will.

Iraq provides an example of not exerting the necessary political will to adopt the modern system. Former Iraqi prime minister Nouri al-Maliki has been criticized for falling back on old patterns of practice, stifling dissent on military decisions, centralizing decision making, and politicizing the officer corps, which prevented Iraq from adopting the institutions that could have saved the Iraqi army from succumbing to ISIS in 2014. On an illustrative basis, at least, this suggests the importance of political will for adopting the system of military reflection.

Developing the political will necessary to reform the military requires a catalyst, an event that triggers a change of direction in society. For example, the Prussians at Jena in 1806 and the French at Sedan in 1870 suffered battlefield defeats that catalyzed society in a new direction. National leaders took advantage of this to push through dramatic, long-lasting military reforms.

A catalyst need not be a direct battlefield defeat, either. The leaders of the United Arab Emirates witnessed the upheaval of the Arab Spring and the subsequent rise of extremism in the region, as well as increased Iranian intervention in the Arab world, and shifted focus on the country's security priorities. The UAE has launched a bid to attain regional power status, and is striving to develop the military institutions to do so. In a similar vein, the military in Brazil was long a tool for internal repression, but a process of electoral change and pressure from civil society in the 1970s and 1980s shifted the discussion on security in the country and gave national leaders the opportunity to make lasting reforms to the military.

In scientific terms, such catalysts are not sufficient, individually, to guarantee the adoption of the system of military reflection, but they do constitute a necessary condition. Whether the catalyst comes on the battlefield or at the ballot box, it is characterized by a shift in the discourse on security, be it to identify the most threatening potential enemies or the best ways to address those threats. In other words, without a catalyst for political will states are unlikely to adopt the system of military reflection.

ACCESS TO CAPITAL

The availability of capital is at least as important as political will. Capital, both human capital and material capital (wealth) provide the means to enact necessary military reforms. Adopting the system of military reflection is a long-term process that requires sustained capital inputs, and many countries have enacted reforms on the heels of a catalyst like a battlefield reversal or a revolutionary societal change, only to find that they could not sustain the reforms. The aforementioned Nizam-e Jedid that the Ottoman Sultan pushed into place in the first quarter of the nineteenth century endured for only a few years, crushed under the weight of mounting external debt and political infighting. Also during the first half of the nineteenth century, the highly literate, centralized Prussian state had a substantial amount of human and material capital to pump into its military reform program over the course of decades, which enabled it to develop staff and educational systems that supported decentralized tactical command and coherence in strategic decision making.

Human capital may be more important than material capital for adopting the system of military reflection. Take the UAE case, for example. Since the mid-1990s, the UAE has been one of the highest defense spenders, per capita, in the world, and the UAE armed forces even deployed to locations as far-flung as Kosovo, Somalia, and Afghanistan. It was not until the first decade of the twenty-first century, however, that the UAE military was recognized as one of the most proficient in the Arab world, one capable of operating alongside advanced Western militaries. This jump in proficiency coincided with an influx of military experts of all stripes coming to the UAE, from training and doctrine departments to military schools. It is also worth noting that the UAE is among the highest in the Arab world in rates of literacy and educational attainment, which suggests that adoption of the system of military reflection cannot simply be bought.

To return to scientific terminology, while having human and material capital available for military reforms is not a necessary condition for adopting the system of military reflection, having *both* the necessary capital and the necessary catalyst for political will are jointly sufficient to ensure adoption. That said, like any human social system, the dynamics of military reform are complex, and I would be loath to reduce this process to one or a few variables, but having political will and available capital dramatically improves the odds of adopting the system of military reflection.

The alternative is disturbing. Militaries that fail to adopt this system of military reflection will be left behind as warfare evolves, and the states they defend will be weak, easy prey to extremists of the most violent order. The reflection system is about the efficient management of violence, and if the first decade and a half of the twenty-first century have shown anything, it is that the security of all will be liable to the strength of a few. In a world of diminishing defense resources, the most efficient use of state violence—in this case, military force—will have an outsized influence on security in the information age.

PLAN FOR THE BOOK

This study uses multiple approaches to investigate how militaries learn. It begins with a statistical analysis of data from over 150 countries since the Napoleonic Wars, simultaneously testing competing explanations for why states adopt the system of military reflection.[6] Using this analysis, the study presents adoption theory to explain the conditions under which states are more likely to adopt the modern system. To increase readability, the more technical aspects of this statistical analysis are in the appendix, although military officers and policymakers reading this book may still find it worthwhile to skip this chapter entirely and move to the case studies that follow.

After identifying the empirical expectations of adoption theory, this study then uses a process tracing approach to test adoption theory on four cases, chosen for their cultural, historical, and political diversity: France, Prussia, Turkey, and Egypt. The final empirical chapter relates adoption theory to an in-depth analysis of military reform in the United Arab Emirates, in particular to the founding and growth of the UAE National Defense College.

It is worth anticipating criticisms that might arise. First and foremost, why is this not simply a study about military professionalism? The answer is that the system of military reflection is a subset of the broad range of characteristics that scholars have associated with military professionalism. Conceptions of military professionalism have become so broad, and vary so much from country to country, that it is necessary to develop a new concept for analyzing how militaries learn. The term "military professionalism" has come to encompass everything from human rights and ethics to military competence and a volunteer force. Using a new term, the system of military reflection, is a useful way of dispensing with excess conceptual baggage.

The second potential criticism is that the conclusion of the study is obvious to a fault. Of course political will and access to capital are necessary for adopting a new set of ideas about how the military should operate. But if the answer is so obvious, why have so many military reform programs struggled to succeed? Clearly, something different needs to be done to increase the prospects of success in advising and military reform.

What is more, the most interesting part of this study is in describing how political will and access to capital have led to adopting the system of military reflection around the world. This is why the study does not stop at establishing statistical correlations, but engages the specifics of military educational reform in a number of countries, including the UAE. There is more than one pathway to military institutional change. Executing a successful military reform strategy depends on accessing capital and political will in a way that addresses the complex needs of culture, politics, religion, and security.

Even if the answer feels obvious, an in-depth analysis of the subject suggests that this is far from the case.

The third potential criticism is that this work has been done before. Two scholarly literatures can plausibly claim to have answered these questions already: the military professionalism literature and the military adaptation literature. However, studies from these two literatures typically focus on one or a few countries, or they do not address the critical factor of how military thought itself changes. In the military professionalism literature, Samuel Huntington describes how military professionalism developed in the United States, and Alfred Stepan offers a pathbreaking work on military intervention in Latin American politics, but neither of these important works extends much beyond their local contexts. In fact, this reluctance to make global claims is fairly widespread in the military professionalism literature. Samuel Finer examines the prevalence of coups, but focuses on the African context, and Peter Feaver focuses on civilian oversight of the military, but only in the United States.[7] In this study, I take the admittedly risky step of making general claims about advanced military education systems, while comparing these claims to case studies in multiple regional contexts. In addition, the case studies range from fine-grained and contemporary (the UAE) to broad and historical (France, Prussia, Turkey, Egypt). In this way, I hope to add a richer general understanding to the more specific, localized observations of the military professionalism literature.

As for the military adaptation literature, it often leaves very interesting parts of the story untold. Peter Katzenstein and Ian Johnston examine the development of military thinking in Japan and China, respectively, leaving unanswered the question of how militaries learn elsewhere. Stephen Biddle argues that force employment—how militaries apply the modern system of military tactics—explains victory and defeat in modern battle, but he admits to not examining the underlying architecture, established during peacetime, that determines whether a military can apply the modern system of tactics during war. I develop the system of military reflection to examine the underlying architecture that Biddle does not. Michael Horowitz, in his *Diffusion of Military Power*, examines the adoption of weapon systems and tactics, such as the aircraft carrier and suicide bombing, but neglects to suggest how the underlying system of thinking changes over time. As Kenneth Pollack suggests in his studies of Arab military effectiveness, simply having a weapon system is not the same as having the intellectual wherewithal to integrate it in the most effective way. In fact, one premise of this study is that the genuine source of military power is not weapon systems and tactics, but the system of military reflection.[8]

Ultimately, the strongest response to criticism is evidence, and my goal in this study is to present evidence that convinces the reader that access to capital and political will are the keys to successfully adopting the system of military reflection. Chapter 1 builds off of the only known large-*n* statistical study of military professionalism to develop adoption theory and the empirical expectations for the case studies in subsequent chapters. Chapter 2 compares two classic cases of military development, France and Prussia, showing how dire military defeats catalyzed the political will to implement long-lasting military reforms. That said, in France and Prussia the development of human capital and the accumulation of wealth did not unambiguously predate military reforms. To address this, chapter 3 examines the case of Turkey, which suffered military defeat after military defeat for generations, but did not begin an era of effective reform in its institutions of military thinking until the middle of the twentieth century, after Turkey had experienced substantial economic modernization, including human capital development. Chapters 1 through 3 thus have the most to say to the military professionalism and military adaptation literatures, in terms of making general statements about how militaries learn.

Chapter 3 also compares Turkey to the case of Egypt, which suffered a dire military defeat in 1967, then began to reform its military. After expelling Soviet advisors in 1972, Egypt turned to the United States. Since 1978, Egypt has sent dozens of officers to American military schools, but most of its military education focus has been on developing local military schools without direct American help, such as the Nasser Higher Military Academy. Both Turkey and Egypt have eschewed foreign help in their military education systems, a slow process that has resulted in fits and starts over the years, but that could ultimately be more sustainable than a quicker approach.

Chapter 4 examines the more rapid approach of the UAE, and digs deeper into the nuance of military learning. Military advisors, especially military advisors to Arab militaries, should find this chapter most interesting. The UAE has sent fewer officers than have Egypt and Turkey to American schools, but it has drawn on much more direct American advice, using American faculty to help establish its National Defense College in 2013. This highlights a contrast between a homegrown strategy and an import strategy for adopting the system of military reflection. In a way, this comparison also evaluates the failed American approach in Afghanistan and Iraq, which focused almost exclusively on training and equipping foreign forces, without considering the societal factors that underlay success.

The concluding chapter sums up the answers to the questions of the study and draws lessons for military advising and security assistance policy. There are two main lessons. First, security assistance is more likely to be effective in the long term if there is a credible system for developing and accessing human capital in the military. Secondly, decision makers should temper expectations of success based on available political will, and should take advantage of crises in security, as in Iraq during its battle with ISIS, to funnel political will toward long-term military reform. The key to success in modern warfare, in both conventional and in (increasingly frequent) complex operations, lies not in the hardware or tactical capabilities a military has, but in its ability to identify, process, and solve security problems holistically using the system of military reflection. The most potent weapons of the future will thus come from the end of a pen, not from the barrel of a gun.

NOTES

1. Posen 1993.
2. Toronto 2011.
3. Waltz 1979; Resende-Santos 1996.
4. Sec, for example, Biddle 2006; Talmadge 2014; Grauer 2016.
5. Gat 2001; Pollack 1996, 2002; Biddle and Long 2004; Biddle 2006; Talmadge 2015; Grauer and Horowitz 2016.
6. For an earlier version of this work, see Toronto 2017.
7. On civil-military relations in the United States, see Huntington 1957; Janowitz 1960; Sarkesian 1981; Feaver 1996, 2009. On the non-American context see Finer 1962; Stepan 1971, 1988; O'Donnell et al. 1986; Nunn 1983, 1992.
8. Pollack 1996, 2002; Katzenstein 1996; Johnston 1998; Biddle 2006; Horowitz 2010.

Acknowledgments

This book germinated as a dissertation then grew into something substantially different, so it has engendered debts by the barnful. I owe some of my greatest debts to the unsung military officers who taught me through word and deed a respect both for professionalism and for the fearsome complexity of the strategic art. Without their unwitting yet earnest contributions over the years, I would have had little reason to finish writing this book, so I hope that my work pays their generosity forward in lives saved, treasure well spent, and peace secured. As SAMSters say, the mind is the key to victory.

I owe a substantial debt to the military officers and government officials I have met and worked with from Middle Eastern countries, whom I could not identify even if I recalled all their names. With each of our interactions, whether they lasted minutes or months, I have tried to watch, listen, and learn. Perhaps, this work will lead to a better understanding of the challenges you face.

This project would not have been possible without the fertile institutional support of the Ohio State University, the US Army School of Advanced Military Studies (SAMS), and the United Arab Emirates National Defense College (NDC). Ohio State provided me a University Fellowship from September 2005 to September 2006, as well as three Professional Enhancement of Graduate Studies (PEGS) Fellowships, stretching from June to December 2004 and June to September 2005. SAMS sponsored me for an Institute of National Security Studies research travel grant to conduct field work for the Egypt chapter, and NDC gave me the test bed to refine my ideas at a late stage of the project.

Portions of this book appeared earlier, in various publications. An earlier version of chapter 1 appeared as "Why Professionalize? Economic Modernization and Military Professionalism," *Foreign Policy Analysis*, vol. 13, no. 4 (October 2017): 854–75. Sections on Turkey appeared earlier as "Training Is Not Enough: Economic Development and Military Professionalism in Turkey," *St. Antony's International Review*, vol. 7, no. 1 (May 2011): 119–43, and are republished with the gracious permission of STAIR. Sections on Egypt appeared as "Active Inaction: Interagency Security Assistance to Egypt," *Interagency Paper* No. 6, Col. Arthur D. Simons Center for the Study of Interagency Cooperation (December 2011). I am grateful to these outlets for allowing me to reprint my work. Such an approach to publishing is the key to a productive marketplace of ideas. I also thank the anonymous reviewers at these publications for their helpful feedback.

A bevy of other individuals enriched this project along the way, so if I forget to mention someone I hope they will accept my deepest apologies. My dissertation committee—John Mueller, Brian Pollins, and Marcus Kurtz—were instrumental in the early stages of the project, and Les Grau and Scott Gorman mentored me at later stages, including commenting on this work. Bryan Batson, Ryan Grauer, and William Shavce read and commented on the complete manuscript, and Daniel Baltrusaitis, Stephen Biddle, Howard Clark, Samuel Greene, Michael Horowitz, Jennifer Jefferis, David Last, Peter Schifferle, and Robert Stewart-Ingersoll read and commented on one or more chapters. Carol Atkinson, Jonathan Canedo, Mary Caprioli, Bridget Coggins, Tom Dolan, Peter Feaver, Paul Fritz, Brian Gaines, Ted Hopf, Pat James, Luke Keele, Ron Krebs, Ed McCaul, Zach Mears, Greg Miller, Irfan Nooruddin, Geoffrey Parker, Randy Schweller, Brent Strathman, Alex Thompson, Daniel Verdier, and Craig Volden read earlier versions of this project. For help of a technical nature, including help with tracking down data, I am grateful to Quintin Beazer, Roman Ivanchenko, David Last, Tamir Libel, David Lincove, Bill Miller, Cory Smidt, and Anand Sokhey. I also thank the anonymous reviewer from Lexington Books for insightful comments on the complete manuscript. A natural outgrowth of these individuals' efforts is an improved work, although they bear none of the blame if the harvest is poor.

My greatest debt is to my family, especially my wife Melissa, who always helps keep my eyes on the prize. This book is dedicated to my children, in the hope that we can all do something great, or at least make the world a little better, or, if not that, then reap what good we sow.

Chapter 1

Adoption Theory

Examining the Intellectual Foundations of Military Success

Scholars and military officers alike spend a great deal of time explaining military success and failure, but more often than not these explanations have focused on tactical competence, regime type, or the preponderance of men or materiel.[1] More rare are explanations that examine the intellectual architecture underlying military success.[2] In relations between states, this intellectual foundation is arguably more important for determining enduring strategic success.

Evidence presented later in this chapter and in the appendix suggests that having a war college is a better predictor of battlefield performance than relative material capabilities or regime type, two factors often identified as correlating with military success. In an age of terror and failing states, it behooves strategists to understand how well military forces build their intellectual foundations. The incoherence of Iraqi operations against ISIS in 2014 reinforces the importance of this "softer" facet of military power. There is more to national power than men, weapons, and democracy.

There are at least two sets of questions involved in the intellectual architecture behind military success. First, some might say that there is only a tenuous link between military education and military performance, so I first show evidence linking the two. Based on my experience observing military forces and conversing with military officers, I use inductive reasoning to suggest why this link between military education and military performance might exist.

The second set of questions seeks to explain why states reform the intellectual architecture upon which military planning and performance is based. I then use statistical evidence to build the empirical expectations behind what I call adoption theory, which links political will, capital, and adoption of the system of military reflection. While subsequent chapters offer historical evidence substantiating the empirical expectations of adoption theory, this chapter starts with the question, "How do militaries learn?"

MILITARY EDUCATION AND BATTLEFIELD SUCCESS

Many militaries educate their officers and NCOs, but detractors often dismiss these programs as having little military utility.[3] Military education courses are generally only given to officers (although the US military often includes NCOs in its courses) and the focus is on "how to think" about military problems, whereas military training generally focuses on the acquisition of specific skills, or on "what to think." In many militaries around the world, officers are educated throughout their careers, usually at four levels: pre-commissioning (academy), junior (in the first five years of service), intermediate (staff college), and senior (war college). These schools often teach a mix of analytical (how to think) and prescriptive (what to think) skills, but as an officer progresses in his or her career, the focus typically becomes more and more analytical.[4]

The question remains, however, whether these analytical skills really matter. Scholars have pointed to unit cohesion, numerical superiority, technology, effective leadership, and ideology to explain battlefield success, but fewer scholars have pointed to military education.[5] Ryan Grauer points to the importance of command structure in explaining battlefield effectiveness, but observes that little is understood about why different command structures are chosen. Military education, or at least the reflection that goes with it, may provide that link.

Critics argue that military education is not essential for motivating a squad of soldiers to clear a house full of booby traps and bad guys, and Dwight Eisenhower did not succeed as commander of allied forces in World War II because he came in first in his class at the US Army Command and General Staff College, but rather because of natural political instincts that would have led to success anyway. While senior military leaders have emphasized the importance of military education, these doubts about military education are a persistent refrain in uniformed and civilian ranks alike.

But my experience teaching and conversing with military officers suggests there are good reasons to believe that military education provides a foundation for battlefield success. The nettlesome environments that militaries face require a daunting command of everything from book-learned knowledge of history and social science to hard-won experience from the world's remotest battlefields and military headquarters. Military officers get this through their education, not only by being exposed to new ideas in the classroom, but also by reflecting on their experience in new ways. Military education becomes a "force multiplier," meaning that it magnifies the positives in what the military is already doing.[6] Military officers who trust in the link between military education and battlefield success often point to how education changes patterns of thinking, allowing officers to analyze with greater precision, and identify creative solutions to unexpected challenges. More military education does not guarantee victory, but it does contribute to enduring military success.

Military education helps establish the conditions for military success by creating the intellectual architecture to make military operations more effective on a continual basis. A brief assessment of modern land warfare suggests why this is the case.

The principal tactical problem of modern land warfare is how to survive the lethality of the battlefield longer than an opposing force. In large-unit land warfare, Stephen Biddle argues that the most effective military organizations survive by employing the modern system of military tactics, which is characterized by effective communication and coordination between different combat arms (such as armor, infantry, artillery, and airpower) in order to maintain the initiative and exploit an opponent's inability to maneuver.[7] The US Army calls this tactical system "combined arms maneuver."[8] When warfare is characterized by insurgency or combat against non-state actors, on the other hand, tactical engagements involve smaller-sized units, and large-unit tactics are counterproductive. Informed by the US Army's "wide area security" concept, counterinsurgents use time between engagements to gather intelligence on enemy combatants, and then use precision munitions or raids to eliminate or capture them. The aim is still to make the battlefield more lethal for foe than friend. Wide area security is still warfare by maneuver, but it is conceived differently than for combined arms maneuver.[9]

How does military education create the "intellectual architecture" to make military organizations better at performing these tasks? I compliment deductive reasoning with inductive theorizing to answer this question, reasoning from experience and refining the theory with statistical inquiry. David Blagden has argued that such an approach has the "truth-preserving quality" of deduction while grounding theories in the real world.[10] So, I build the theory with evidence, after launching from observed experience.

In my experience, militaries build the intellectual foundations of battlefield success by aspiring to excellence. When military education teaches critical thinking and exposes practitioners to theory, it contributes to excellence throughout a military organization. Over the years, I have discussed assessments of foreign militaries with dozens of American and allied military officers, and this notion of excellence is a consistent theme. My informal discussions with military officers and those who teach them suggest three characteristics of military education that lead to military excellence.

The first characteristic is stable civil-military relations, and civil-military relations that allow space for the military to develop its expertise, providing the military a well-defined role in setting security policy and establishing spending priorities.[11] For example, the US Department of Defense submits budget requests to the White House, but does not act on those spending priorities without Congressional approval. Elements of the US military, both individual officers and Pentagon officials, contribute to a vibrant discussion on the nation's security policies, at least those policies related to the military's mission. What is more, the US military has very wide latitude in training and equipping its forces. As an element of civil-military relations, this latitude is just as important as stability in contributing to military excellence.

But this is not the only characteristic of military excellence. Institutional introspection also plays a role. In the United Kingdom, this institutional introspection often plays out in formal legislative oversight, where the military publicly evaluates its own performance, but this institutional introspection also percolates from national, strategic spheres down to tactical ones. For example, the US Army has a Center for Army Lessons Learned that collects and disseminates lessons that soldiers learn from the battlefield, and junior officers also use resources like *War on the Rocks* and *Small Wars Journal* to share lessons from combat. A reflective military looks at itself and its performance, and often undertakes rigorous training regimes to consistently improve its competence, including exercises that practice integration between tactical and strategic decision makers, and exercises that operate on a twenty-four hour basis instead of just eight hours per day. Having institutional introspection does not mean that military operations will result in unmitigated success, but it does mean that a military can learn from its mistakes and train itself to avoid them in the future.[12]

In addition to having stable civil-military relations and institutional introspection, militaries with an aspiration to excellence also cultivate a capacity for critical analysis. In fact, the most effective militaries seem to integrate critical analysis into school curricula and planning processes. Beginning prior to World War II, for example, the UK military developed its "estimate of the situation" at all levels of military planning; British officers are now well-versed in the "seven questions" they are to ask while planning military operations. The US Army has its "design methodology" and the "military decision-making process," which fulfill similar functions.[13] In my years teaching military officers, I have observed that the hardest-hitting questions in classroom discussions and after lectures often come from officers who have served in combat. During the years I taught for the US Army, when tanks were rarely used in conventional combat roles, armor officers would consistently rail against the inability of crews to meet the tank gunnery standards necessary to be effective in combined arms maneuver warfare.

The complainants were not disgruntled officers stepping on a soapbox, but rather interested observers critically analyzing elements of warfare that the military was not then participating in. This critical thinking goes hand in hand with military excellence.

So, military excellence—stable civil-military relations, institutional introspection, and a capacity for critical analysis—is the conceptual link between military education and military effectiveness. Biddle does not specify the conditions and causes of adopting the modern system of tactics, nor does Grauer clarify when militaries will adopt a tactically appropriate command structure, but I contend that actively cultivating military excellence is the central component of the system of military reflection. While doing so is no guarantee, militaries that adopt the system of military reflection are more likely to produce military success, because it engenders excellence at every level of a force.

History presents important challenges to this notion that the system of military reflection is the link between military education and battlefield success. For example, the German army was regarded as one of the most proficient military organizations in the seventy years after the Franco-Prussian War, but German strategic blunders undermined victory in two world wars. The US military has arguably been the most effective battlefield force in the post-9/11 world,[14] but this has not translated into strategic success in Iraq and Afghanistan. What is the point of being militarily effective if it does not result in political victory? The answer is that war is the province of passion and chance, as well as reason,[15] and that military effectiveness is no guarantee of political success. However, states have a harder time achieving political success when they do not have effective militaries.[16] Why else would states spend so much time and effort to man, train, educate, and equip military forces?

Empirical evidence suggests that all this time and effort is actually good for something. Ryan Grauer and Michael Horowitz, in one of the most recent examinations of military performance, use a dataset of decisive operations from all interstate wars between 1913 and 2003 to correlate battlefield effectiveness and Biddle's concept of force employment. Grauer and Horowitz argue that forces implementing the modern system of tactics and operations were systematically more likely to win in eighty-three of these decisive military engagements.[17]

Table 1.1 reports my replication of Grauer and Horowitz' main result, then adds two new variables: whether the participants in the eighty-three engagements had a staff college (or a war college) at the start of the engagement. Two observations are of note. First, Grauer and Horowitz' main finding regarding force employment and battlefield effectiveness is largely preserved, despite the addition of the two military school variables. The adoption of

Table 1.1 Military Education and Battlefield Victory, 1913–2003

	Grauer & Horowitz Probit Model		
DV= Tactical victory	*Replication*	*w/Staff College*	*w/War College*
Staff college		**1.993**	
War college			**4.378**
Comparative modern system adoption level	**1.442**	**1.683**	**2.919**
Polity score	**–.644**	**–.880**	**–.859**
Polity squared	**.026**	**.033**	**.024**
Polity side B score	.347	.478	**.813**
Polity side B squared	–.019	**–.027**	**–.043**
Comparative military capabilities	**2.333**	**2.236**	1.806
Strategy (maneuver/attrition/punishment)	.333	**.878**	**1.043**
Terrain	3.483	**5.429**	**8.190**
Comparative troops engaged	.566	–1.405	–3.025
Comparative human capital	**.002**	**.003**	**.005**
Comparative civil-military relations	–.205	–.307	–.907
Constant	**–4.644**	**–6.799**	**–10.013**
N (observations)	83	83	83
N (countries)	36	36	36
Pseudo-R^2	.755	.799	.840
Probability>χ^2	.000	.000	.000

Note: Significant coefficients in **bold** ($p<.05$, two-tailed); robust standard errors clustered on country code.

the reflection system still correlates with tactical victory. Secondly, the two models that include information on the presence or absence of military schools explain more of the variation in battlefield performance than Grauer and Horowitz' original model, as the pseudo-R^2 statistic indicates. The staff college model explains almost 5 percent more tactical wins and losses than the original model (.799 as compared to .755), while the war college model explains over 8 percent more of the tactical outcomes than Grauer and Horowitz' original model.

Replicating Grauer and Horowitz' results constrained the analysis to the 83 engagements for which data are available, but their full data set contains force employment data on 125 engagements. In an analysis of the full data set, described fully in the appendix, I correlate my data on military schools with Grauer and Horowitz' data on the extent to which the forces engaged employed their forces according to the modern system of tactics and operations. Countries with a war college are 19 percent more likely to implement the modern system of tactics fully and 18 percent less likely to not implement it at all in these 125 engagements.

Even though this is one of the most complete and valid datasets reported in the military effectiveness literature, winning a tactical engagement is not the same as winning a military conflict, so I conduct robustness checks using data on 140 interstate wars from 1945 to 2003 (from the Correlates of War Project). I find that countries with a war college are 20 percent more likely to be victorious in interstate wars than those without a war college. In a much larger sample of observations—5,321 militarized interstate disputes from 1816 to 2011, which includes actions ranging from the threat of military force to interstate war—states with a war college also demonstrate greater military effectiveness. They are 5.9 percent more likely to suffer no fatalities during such disputes, and 1.7 percent less likely to suffer 1,000 or more fatalities during such an interstate dispute. So, having a war college corresponds with suffering fewer casualties during disputes, winning wars, winning battles, and implementing the modern system as defined by Biddle. Military excellence, both tactical and intellectual, matters, which suggests that adopting the system of military reflection contributes to battlefield success, and that studying how militaries learn is a worthwhile endeavor.

BUILDING THE SYSTEM OF MILITARY REFLECTION

If the system of military reflection contributes to battlefield success, then how do states adopt it? Most militaries develop a capacity for critical analysis and institutional introspection through military education systems, but societal norms and national myths also play a role. For example, Israeli society has a strong preference for speaking one's mind and engaging in vibrant debate, and national leaders perpetuate myths of threat that push all members of society to excel at what they do. This has allowed the Israel Defense Force to create an aspiration to excellence without a heavy reliance on military education.

That said, the dominant pattern since 1800 has been for states to adopt the reflection system by reforming military education and associated military institutions. The timing and nature of this commitment to military education have varied from state to state, but almost no studies have examined this question from a large-*n* perspective. There are many qualitative or theoretical studies dealing with military professionalization, and there are a number of policy documents that emphasize the importance of training foreign militaries, but so far little of the conventional wisdom on military educational reform is informed by analysis that examines a large number of countries over a long period of time.[18]

This is especially puzzling in the post-Cold War world, when complex security problems have multiplied and defense spending has shrunk. Now, more than ever, states and societies should be concerned with the efficient use of military resources, instead of simply doing things the same old way. The information age makes the adverse consequences of military operations acute, since enemies can exploit success through information. What is more, the traditional focus on in-depth case study analysis views military reform through the lens of political culture and regime type, or of security concerns.[19]

Examining reform in military education from a large-*n* perspective gives a different picture of military educational reform. This study analyzes data on 156 countries from 1816 to 2011. These data suggest a two-step process: states with more severe security concerns are more likely to reform military education, but states with a higher level of economic modernization, as measured by wealth and human capital, are more likely to maintain these reforms over the long term. This two-step process explains reforms in military education better than regime type, security concerns (in isolation), or colonial heritage, factors that scholars have suggested previously may account for changes in military institutions.[20] Thus, the path to adopting the system of military reflection over the long term, the sine qua non of future security in the twenty-first century, is less about democratization and more about the internal economic processes associated with modernization, such as acquiring and employing human capital and wealth.

Reforming Military Education

Scholars of military professionalization and military adaptation occasionally discuss military reform, but these literatures have often focused on conflict between civilian and military leaders, on the ability of the state to manage the use of violence, or on military interventions in politics.[21] It is more rare for scholars to examine reform in military education in particular.[22] Many studies of military institutions also focus on one or a few countries, or on describing how these institutions change instead of why. Other studies give cultural or institutional explanations for why military institutions change, but they tend to be based on a very small number of cases, which makes them difficult to generalize or, if generalization is possible, difficult to measure.[23]

This study contributes to these literatures by focusing on why militaries change the way they learn. Militaries often adapt to conditions during war, but practitioners and observers of strategic affairs should be more concerned with the conditions under which states will engage in long-term reform of military education, since it is this type of fundamental change that leads to enduring advantage on the battlefield. Like professionalizing the military, reforming military education is about changing the way the state manages violence, since it presupposes changes in who can appropriately use violence in society, how they do so, and to what end. Military educational reform, then, can be considered a subset of military professionalization and military adaptation, which is why I examine these literatures here, asking how they would answer the question, "Why do militaries change the way they learn?" Previous scholarship suggests four main answers to this question: regime type, international security, economic modernization, and culture.

Regime Type

Many of the classic studies of civil-military relations characterize the management of military violence in terms of democratic transitions or of the retreat to the barracks and civilian control.[24] Applied to reforming military education, this literature would expect militaries to change the way they learn as society democratizes, since democracies have less conflictual civil-military relations and tend to have robust, liberal control over the military, making it easier to overcome institutional inertia against reform and to improve more quickly in the face of poor military performance.

More recent scholarship has refined this idea. Dan Reiter and Allan Stam examine military effectiveness, arguing that democratic militaries are more loyal to the regime, more efficiently organized, and more meritocratic. Stephen Biddle and Steven Long argue that "superior human capital, harmonious civil-military relations, and Western cultural background" explain why democracies are so successful in war, although they admit that it is just as likely that democracy is caused by these factors as that it causes them.[25]

In general, the expectation of the regime type literature is that democracy will be positively associated with reforms in military education:

Regime Type Hypothesis: Democratic states are more likely to reform military education.

International Security

A second literature would argue that international security concerns explain why some countries reform military education and set themselves on the pathway to adopting the system of military reflection. International security concerns contributed to the rise of professional military institutions in Germany and France, and Eliot Cohen finds that security concerns are an important consideration in the determination of military manpower systems. Michael Barnett finds that states use participation in war to ratchet up the role of the state in society. Barton Hacker argues that technological innovations in the nineteenth century drove military modernization and reform; if international security influences the timing of adopting these technological innovations, then it would explain differences in the timing of military educational reform.[26]

A great deal of qualitative work also links international security concerns to military educational reform, especially in South American and European countries. For example, Argentina, Brazil, Chile, and Peru chose between the French and Prussian military models on the heels of the Franco-Prussian War and the Wars of the Pacific, choices that influenced patterns of civil-military relations in these countries in the twentieth century.[27] Furthermore, the neorealist tradition in international relations theory contends that security concerns are key to socializing states to the military institutions of the day. João Resende-Santos, for example, sees higher levels of military reform around the world on the heels of great power war, when some military institutions succeed and some fail on the world stage and states scramble to adopt the most successful patterns of practice.[28] This reasoning suggests the International Security Hypothesis:

International Security Hypothesis: States with greater security concerns are more likely to reform military education.

Culture

Scholars have also noted the importance of culture and institutions on the development of the military.[29] Some of this scholarship focuses on military reform. Theo Farrell asks how newly democratic militaries adopt norms of civilian supremacy and conventional warfare. He finds that Ireland relied on a "process of political mobilization whereby the target community is pressured into adopting new ways of thinking and doing."[30] Derek Lutterbeck explains the variation in how Arab militaries responded to revolts beginning in 2011 by pointing to "the degree of institutionalization of the armed forces and their relationship to society at large," while Robert Springborg and Henry Clement point to "social location" to explain why the Egyptian and Tunisian militaries behaved differently during these revolts. Kenneth Pollack and Biddle and Long identify culture as a factor in military effectiveness. In research published with others, I have found a link between colonial heritage and military recruitment choices; former British colonies are about half as likely to use conscripts as those countries without a British colonial heritage.[31]

This scholarship suggests a concept of "cultural distance" when it comes to military educational reform. In short, modes of thinking inherited from the cultural milieu may necessitate a deeper level of reform in order to adopt the most effective institutions in military education. For example, former colonies often model their military education systems on those of former colonial rulers, so the relationship between former colony and colonizer, be it conflictual or amicable, may influence the timing and depth of reform. Countries in which an Indo-European language dominates may be more likely to adopt the system of military reflection, as it more closely resembles everyday modes of thinking as compared to countries where an Indo-European language is used only on a secondary or tertiary basis. In the terms presented in this study, if a culture has less tolerance for dissent from authority and criticism of others, or if face-saving plays a prominent role in daily life, then institutional introspection and a capacity for critical analysis will be more difficult to develop, and so it will be more difficult to adopt the system of military reflection. Assuming that these dynamics are distinguishable from regime type, international security, and economic modernization, then the following hypothesis should hold:

> *Cultural Distance Hypothesis:* States with more cultural distance to cross to implement reforms are less likely to reform military education.

Economic Modernization

A fourth body of literature would contend that states are more likely to reform military education if the economy has modernized significantly. The idea here is that an advanced military education system is an important marker of a modern society, one that suggests changed attitudes about the role of the military in the management of violence.[32] As society passes from traditional to modern, institutions take on characteristics of impersonal authority, standard operating procedure, and striving for efficiency, which can foment an aspiration to excellence in a country's military force.[33]

But the effect of modernization on adopting the system of military reflection is more direct than this. Two pathways link economic modernization and reforms in military education, which can in turn lead to adopting the reflection system over the long term. First, a modern economy implies greater wealth, from which state leaders can draw to underwrite reforms. Preparing the military for war is not directly productive economically. The military provides security, but generally does not add to the value of goods and services by increasing efficiency, and professional militaries are generally staffed by career-service officers and men, not conscripts. Professional armies are more expensive to maintain on a per soldier basis. Reforming military education thus costs a lot of money, especially over the long term, so it requires an accumulation of material capital.[34] This suggests the Wealth Hypothesis:

> *Wealth Hypothesis:* Wealthier states are more likely to reform military education.

Human Capital

The accumulation of human capital is the second pathway between economic modernization and an advanced military education system. The professional military officer may be a specialist in the management of violence, but this does not mean that an officer's knowledge is limited to this one field. The most successful officers are also generalists who appreciate the multifaceted implications of military theory, politics, economics, science, history, and even literature on the conduct of military operations. For example, it is not uncommon in the US officer corps today to find individuals with multiple Master's degrees, or to find undergraduate majors as diverse as political science and chemical engineering represented. Most of the Western military officers that I have met seek out an impressive variety of knowledge. They recognize that the exact parameters of military problems are unknown before they occur, and they recognize the importance of working with others in order to solve complex problems.[35] American officers in Afghanistan have noted the importance of developing human capital in society in order to develop the country's military force.[36] Thus, human capital development is at the core of an advanced military education system, contributing to a capacity for critical analysis, institutional introspection, and stable civil-military relations, hallmarks of the reflection system. This is the thinking behind the Human Capital Hypothesis:

> *Human Capital Hypothesis:* States with a higher level of human capital are more likely to reform military education.

Economic Growth and Military Reform

There is a possibility that economic growth could be the result of military reform processes, not the other way around. As one scholar put it to me, "Suspicion lurks that the causality is reversed." By this reasoning, as military institutions become more rational and efficient, the state's application of violence also becomes more predictable. This predictability is important for economic development, as it presents conditions ripe for successful entrepreneurship. As long as the costs of doing business are predictable, however high these costs are in an objective sense, then entrepreneurs can make a profit by pursuing the activity that offers them the lowest opportunity cost. In this way, military reform would lead to economic modernization.

Not all evidence supports this contention, however. While the military in some countries began modernizing reforms prior to the outbreak of economic development—Prussia is a prototypical example—there are just as many cases where militaries professionalized either in conjunction with or following economic development. France, for example, radically reformed its military on the heels of the Franco-Prussian War, which occurred after thirty years of remarkable French economic progress. In Brazil, professional military institutions developed at the same time that the economy developed in the first half of the twentieth century.[37]

A close examination of the evidence suggests that the process of adopting the system of military reflection is more complex than that. Neither does military reform cause economic modernization, nor vice versa. Instead, cues in the security environment (poor battlefield performance, rivalry with another state, a quest for regional power status) catalyze the political will necessary to launch reforms, but the prior extent of economic modernization (as indicated by the accumulation of wealth and human capital) influences adoption over the long term.

STATISTICAL EVIDENCE

More specific data analysis can be found in the appendix, but the evidence suggests that the security environment interacts with the level of economic modernization to explain the timing and extent of military educational reform. I examine four indicators to assess when and how countries begin to adopt the system of military reflection: the establishment of a staff college, the establishment of a war college, the number of military periodicals, and military expenditures per soldier. I follow the lead of scholars who have used military expenditures per soldier to measure martial quality,[38] but the other three indicators represent original data collection.

In previous research, I used the military periodicals data and data on whether countries had a military academy to educate officer cadets to measure the extent of military professionalism,[39] but I found that pre-commissioning academies are so widespread around the world that they do not meaningfully discriminate between countries. The presence of staff colleges, however, and especially of war colleges, are much more meaningful discriminators. In fact, today only sixty-three countries have a war college (sometimes termed a defense college). What is more, whereas academies tend to focus on imparting skills that cadets will need in the first few years of their career (how to lead small groups and train tactical proficiency), officers with ten to fifteen years' experience begin to address the complexity of warfare during their staff college education, in particular how to align military power in time, space, and purpose to achieve strategic objectives. War college education is even more focused on complex problems, as students must grapple with how to coordinate all elements of national power (including nonmilitary components like the economy and diplomacy) in order to implement national strategies and preserve national interests.[40] Having staff and war colleges, then, is an important marker of adopting the system of military reflection.

Table 1.2 reports statistical results that explain the timing of establishing staff colleges and war colleges in over 150 countries. In technical terms, these are Cox survival models, which means that, in each year, they compare all the characteristics of countries that have not yet established a military school with the characteristics of countries that establish these schools in that year, to see where they differ.[41] *GDP per capita* is included to measure wealth (and test the Wealth Hypothesis), whereas *democracy* is included to examine the Regime Type Hypothesis. To test the International Security Hypothesis, I include information on whether a state is involved in an *interstate rivalry* with any other state; whether it is fighting or has fought an interstate war in the previous five years (*recent war participation*); whether the world is currently or in the last five years has experienced a war between any two or more great powers (*recent great power war*); whether a state is itself a *great power* (since they have a much wider variety of security challenges than other states); and an index of a state's *material capabilities* (total and urban population, energy consumption, iron and steel production, military expenditures, and size of the military).

Table 1.2 Establishing Military Schools, 1816–2011

Explained Variable →	*War College Established*	*Staff College Established*
GDP per capita	.789	1.097
Interstate rivalry	1.577	**1.774**
Recent war participation	1.161	1.231
Recent great power war	**2.782**	1.745
Great power	.693	.823
Material capabilities	**6.67×e[10]**	**3.7×e[6]**
Democracy	1.686	.916
Colonial ruler model	**.984**	**.986**
Violent independence	.961	1.351
Col model×Violent ind	1.011	1.007
Time since independence	**1.010**	.999
N (observations)	8,408	6,002
N (countries)	154	150
Probability>χ^2	.000	.000

Note: Significant coefficients in **bold** (*p*<.05, two-tailed), using Cox survival models.

I include three variables to begin to examine the Cultural Distance Hypothesis. *Colonial ruler model* equals 0 until a state's former colonial ruler establishes a war college (or a staff college, for the staff college model), after which it begins counting years since the former colonial ruler established the military school. The reasoning is that former colonial ties engender a cultural connection—the longer a state goes without establishing a military school, while its former colonizer has, the more likely that state is to follow suit, since it must travel less cultural distance to mimic the former colonizer than to mimic other potential models. If a state has undergone a *violent independence* from the colonial ruler, however, I expect this logic to be reversed—a state would go to great pains not to mimic its former colonizer. The interaction term between these two variables (*col model×violent ind*) means that, if independence was violent, the more time passes after a former colonizer establishes a military school, the less likely the colony is to do so. Thus, the Cultural Distance Hypothesis would expect the coefficients for this variable to be negative. By the same token, because this interaction term is included, the coefficients for the *colonial ruler model* represent only those states for which independence was peaceful, so the Cultural Distance Hypothesis suggests that these coefficients should be positive.

I include *time since independence* because previous research has shown this to be an important factor in state development, including the development of military institutions.[42]

These results suggest some support for the International Security and Cultural Distance Hypotheses and no support for the Regime Type and Wealth Hypotheses. States engaged in a rivalry are more likely to establish staff colleges, while recent great power wars are associated with the establishment of war colleges. The extent of a state's material capabilities is correlated with the opening of both types of school. That said, the level of economic wealth seems to play no role in the timing of establishing military schools.

But establishing military schools is not the same as maintaining them over the long term. Table 1.3 presents results that examine this slightly different question. These models include the same variables as in Table 1.2, although instead of examining why states establish military schools in the first place, the data is set up to examine the probability that a state will have a staff college or war college in any given year. The models in Table 1.3 also examine why the number of military periodicals published is greater in some states than others, as well as what explains a higher level of military expenditures per soldier.

Only the Wealth and Cultural Distance Hypotheses receive consistent support across all four indicators of maintaining the system of military reflection, although *material capabilities*, meant to examine the International Security Hypothesis, is significant and in the expected direction for the three models it is appropriate to include it in. None of the other factors linked with the International Security or Regime Type Hypotheses has the expected effect across all four models. Being a democracy, having an interstate rivalry, being a great power, and having large material capabilities are associated with only two or three of the four indicators of the system of military reflection. Sometimes these indicators go in the opposite direction from what the hypotheses would expect.

Table 1.3 Maintaining the System of Military Reflection, 1816–2011

Explained Variable →	*War College*	*Staff College*	*Military Periodicals*	*Military Expenditure per Soldier*
GDP per capita	**.165**	**.401**	**.674**	**1.062**
Interstate rivalry	**1.166**	**.336**	.212	.234
Great power	**.330**	**–.334**	.369	**–1.117**
Material capabilities	**4.598**	**11.786**	**7.952**	
Recent war participation	**.622**	**.517**	**.291**	.151
Recent great power war	**–1.469**	.022	–.008	**–.832**
Democracy	**.824**	**–.155**	**.546**	.096
Colonial ruler model	**.006**	**.007**	**.002**	**.324**
Violent independence	**.193**	**.554**	–.252	–.553
Col model×Viol indep	**–.009**	–.001	.001	–.010
Time since independence	**.014**	.002	.011	–.008
Constant	**–5.427**	**–6.401**	**–6.412**	**–2.570**
N (observations)	9,854	9,854	9,710	7,136
N (countries)	156	156	156	156
R^2 or pseudo-R^2	.324	.195	.225	.572
Probability>χ^2 or F	.000	.000	.000	.000

Note: Significant coefficients in **bold** (*p*<.05, two-tailed); robust standard errors; coefficients for time controls not reported; war college and staff college models: logit; military periodicals model: negative binomial; military expenditures per soldier model: ordinary least squares.

The level of economic wealth, though, is associated with maintaining a staff college and war college, with publishing more military periodicals, and with spending more per soldier. For example, wealthy states are 2.3 percent more likely to have a war college than poorer states.[43] This is likewise true for those states that experienced peaceful independence from a colonizer. States that experienced peaceful independence and whose former colonizer has had a war college for twenty-five years (the mean) are 0.9 percent more likely also to have a war college, whereas similar states with a violent independence experience are 0.4 percent less likely to have a war college. States with greater material capabilities are 1.2 percent more likely to have a war college. In comparison with competing factors, then, wealth has a substantial effect on the probability of having a war college.

The evidence presented thus far has not examined the Human Capital Hypothesis, because data to measure the accumulation of human capital were not readily available for a large number of countries prior to 1970. Given this, I also report results in Table 1.4, which replicates the models from the previous table using indicators of human capital accumulation instead of wealth accumulation. In this case, I use the proportion of first grade-age children that actually enroll in first grade (*grade 1 intake*) and the expected number of years a student will attend college (*expected years of college*) to measure the level of human capital in society. Both indicators have good data coverage, and a factor loading analysis indicated that they were sufficiently distinct from one another to be used in the same statistical model. While using these indicators constrains the sample to the years from 1970 to 2011, it allows me to test the Human Capital Hypothesis directly.

The results suggest strong support for the Human Capital Hypothesis, at least in years since 1970. Both human capital indicators are consistently associated with all four indicators of the system of military reflection. On the other hand, these data show only inconsistent support for the International Security, Regime Type, and Cultural Distance Hypotheses. None of the indicators for these three hypotheses show a consistent relationship across all four models.

Figure 1.1 illustrates the importance of both wealth and human capital in adopting the system of military reflection. Taking a snapshot in time (2010, the last year for which data are available), this graph plots all countries as a function of wealth (GDP per capita) and human capital (expected years of college education), then indicates which of those countries have war colleges and which do not. The results indicate that countries with war colleges are more likely to be wealthy and enjoy high levels of human capital, which corresponds with the statistical data already presented. The line in the figure represents the effect of GDP per capita on the probability of having a war college; substantively, a one standard deviation increase in wealth translates

Table 1.4 Human Capital and the System of Military Reflection, 1970–2011

Explained Variable →	*War College*	*Staff College*	*Military Periodicals*	*Military Expenditure per Soldier*
Grade 1 intake	**.016**	**.007**	**.017**	**.007**
Expected years of college	**.555**	**.696**	**.577**	**.688**
Interstate rivalry	.230	.085	.096	–.108
Great power	.884	**2.028**	.562	**.817**
Recent war participation	–.109	**.901**	.208	.065
Material capabilities	**148.928**	**39.046**	**38.334**	
Democracy	–.048	**–.575**	–.151	.216
Colonial ruler model	**-.042**	**.005**	.001	.081
Violent independence	**-1.077**	.377	–.108	.093
Col ruler×Violent ind	**.026**	.002	–.001	–.041
Time since independence	–.004	**–.009**	**.016**	.001
Constant	**–2.950**	**–5.392**	**–3.277**	**7.282**
N (observations)	2,905	2,905	2,880	2,315
N (countries)	151	151	151	148
R^2 or pseudo-R^2	.495	.306	.235	.387
Probability>χ^2 or F	.000	.000	.000	.000

Note: Significant coefficients in **bold** (p<.05, two-tailed); robust standard errors; coefficients for time controls not reported; war college and staff college models: logit; military periodicals model: negative binomial; military expenditures per soldier model: ordinary least squares.

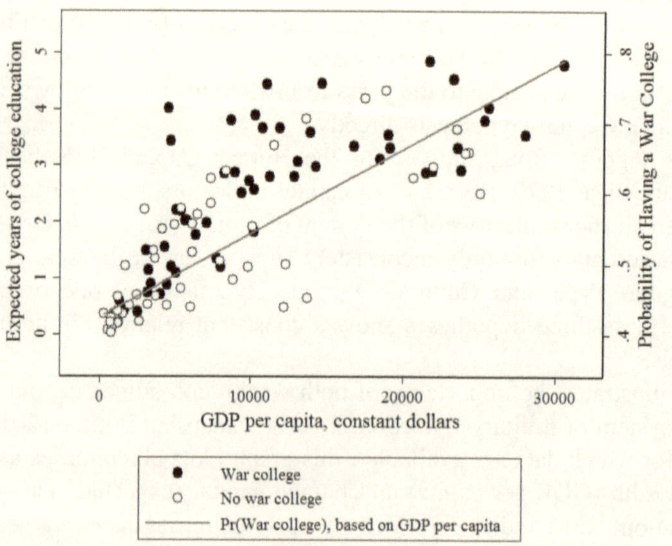

Note: GDP data from Maddison Project 2013; education data from World Bank 2016.

Figure 1.1 Human Capital, Wealth, and War Colleges (All Countries, 2010). *Source: GDP data from Maddison Project 2013; education data from World Bank 2016.*

into a 10.5 percent increase in the probability that a country will have a war college.[44] Over the last two hundred years, and especially since 1970, economic modernization has been the key to maintaining the system of military reflection.

ADOPTION THEORY

The evidence presented here suggests two stages in the process of adopting the system of military reflection. In the first stage, states take a cue from the international security environment in order to launch military reforms. The security environment generates the political will necessary to begin reforms, presumably in the form of a battlefield catalyst, but the statistical evidence presented above suggests that this catalyst is not enough to maintain the system of military reflection over the long term. For that, the second stage is necessary: economic modernization, or the constant accumulation of wealth and human capital.

What effect do battlefield catalysts and economic modernization have on the process of adopting the system of military reflection? In a large-*n* statistical framework, it is very difficult to see nuance in the adoption process, much less formulate different ways in which adoption could unfold, since statistics isolate the mean effect of a set of explanatory variables on one variable to be explained. The case studies that follow illustrate this nuance in a much richer way than the statistics can.

Figure 1.2 gives the theoretical expectations of adoption theory. In general, adoption theory is concerned only with those states that have demonstrated a certain determination to reform the military and, hopefully, adopt the system of military reflection. It is less concerned with states like Micronesia and Panama, which do not have militaries, or Barbados, which has a military but whose environment will likely never require it to significantly improve upon the security that its 500-man force can provide. Adoption theory is likewise unconcerned with great powers that have clearly adopted the system of military reflection, like the United States, the United Kingdom, and China. Rather, adoption theory's sweet spot is in understanding all those states in the middle, states that will shape local and regional security in the twenty-first century.

This 2×2 table identifies four likely pathways to adoption, based on the battlefield catalyst and economic modernization. If a state has already accumulated a substantial amount of wealth and human capital when it experiences a clear battlefield defeat, then adoption is likely to be radical. This is because the battlefield defeat will have generated a substantial amount of political will, and the raw material to make radical changes in how the

Battlefield Catalyst	Economic Modernization: Modern	Economic Modernization: Progressing
Clear Defeat	*Radical Adoption* (France after 1871)	*Front-loaded Adoption* (Prussia after 1806, Egypt after 1967)
External Pressure	*Steady Adoption* (UAE after 2011)	*Fitful Adoption* (Turkey after 1796)

Figure 1.2 Expectations of Adoption Theory. *Author.*

military operates (wealth and human capital) will already be in place. France after 1871 (chapter 2) is an example of Radical Adoption. In this case, military reform was swift and complete.

If a state has not achieved a substantial level of economic modernization prior to experiencing a clear battlefield defeat, then adoption will likely be less radical, although still front-loaded. Military reform in the Front-loaded Adoption pathway is substantial, though not as deep and far-reaching as with Radical Adoption. The reason is that, while the political will to reform the military exists, the wherewithal to continue reforms only becomes available as the economy modernizes, as happened in Prussia after 1806 (chapter 2) and Egypt after 1967 (chapter 3). In these two cases, a period of steady reform followed substantial initial changes in how the military operated.

At times, the battlefield catalyst comes in the form of constant external pressure rather than a clear battlefield defeat. In this case, if a state has not achieved a substantial level of economic modernization, then the likely pathway is fitful adoption, as in Turkey after 1796 (chapter 3). Cases of fitful adoption enjoy neither the wealth and human capital nor the political will necessary to ensure deep and immediate adoption of the system of military reflection. Instead, they may implement some reforms, but then regress when the resources to underwrite reforms fall short, or when conservative forces decry the threat to traditional values that reforms entail. In a situation of constant external pressure and incipient economic modernization, the forces of political will to launch reforms and lack of material and human capital resources to see those reforms through push and pull countries to and from adoption of the system of military reflection on a cyclical basis.

For countries that have substantially modernized their economies prior to experiencing constant external pressure, the pathway to adoption of the system of military reflection will resemble a gradual upward slope, what I call Steady Adoption. In these cases, external pressure will take the form of a quest for regional power status or an enduring rivalry with another state, but will stop short of a dramatic battlefield defeat whose effects are felt throughout society. Thus, the international security environment does not present overwhelming cues in favor of reform, although leaders may recognize the need for improving military effectiveness. At the same time, state leaders have not only the wealth but the human capital to make these reforms stick, and the political space to fend off, if need be, conservative voices raised against gradual reforms. For example, the United States gradually adopted the system of military reflection in response to criticism of its military performance in the Spanish–American War and World War I, leading it to create the force that led the allied offensives on the Western Front in World War II and helped establish the United States as a superpower after the war.[45] The United Arab Emirates after 2011 (chapter 4) may be another example of Steady Adoption. In the face of increasing Iranian interventionism, it launched a bid for regional influence in the aftermath of the Arab Spring, but while it enjoys tremendous material wealth, it also lacks a deep reserve of domestic human capital, which presents interesting possibilities for drawing conclusions about the role of human capital in adopting the system of military reflection.

The case studies that follow employ a process-tracing approach to draw out the nuance in these four pathways to adoption.[46] A large-*n* statistical analysis can identify overarching trends over time, such as the two-stage process linking international security concerns and economic modernization to adoption of the system of military reflection. Adoption theory goes beyond this initial observation to generate specific expectations that, if verified in the empirical record, would substantiate adoption theory.

The cases themselves—Prussia after 1806, France after 1871, Turkey after 1796, Egypt after 1967, and the UAE after 2011—are chosen with an eye toward regional and methodological diversity. Prussia and France are prototypical examples of European military reform,[47] and they are included here to reinterpret the pattern of reform in light of adoption theory. The other three cases are Middle Eastern, a region where scholars have noted a lack of both state development and military effectiveness.[48] As such, these are hard cases for adoption theory, since factors besides a battlefield catalyst and economic modernization may militate against adoption.

If, however, these cases conform to the expectations of adoption theory, as outlined above, then not only will it strengthen adoption theory but it will also suggest that the Middle East region is not as unique as scholars have claimed. For example, scholars have studied the development of military institutions in Egypt and Turkey extensively,[49] but never cross-regionally or in the context of a global theory of how militaries learn. This new context may begin a reappraisal of how military institutions work in the region.

The UAE case study is diverse from a methodological standpoint, since it breaks from the historical approach, based on secondary sources, and follows a more anthropological approach. As of this writing, I am a professor at the UAE National Defense College, so I have witnessed the process of adoption firsthand in the UAE. While I am invested in the process itself and have a stake in its success, this approach also gives a level of nuance that a strictly objective social scientist, making observations from a distance, cannot achieve. If the expectations of adoption theory generated in the statistical analysis are upheld in this in-depth, anthropological study of the UAE, then it would doubly strengthen the theory and, potentially, provide a model for methodological diversity in other social scientific studies of state development and military effectiveness.

On the question of how militaries learn, a clear two-stage pattern emerges from the historical record: international security concerns catalyze reform, but economic modernization makes it a reality. Adoption theory suggests how these factors interact, generating four different pathways to adopting the system of military reflection: radical, front-loaded, fitful, or steady. The case studies that follow illustrate these patterns and highlight the role of human capital in the adoption process. States that aspire to military excellence cannot rely only on security concerns to motivate adoption of the system of military reflection. They also need modern economies.

NOTES

1. Reiter and Stam 1998; Desch 2002, 2010; Biddle and Long 2004; Biddle 2006; Gartzke 2007; Ramsay 2008; Beckley 2010; Renshon and Spirling 2015; Grauer and Horowitz 2012; Grauer 2016. Biddle 2006 and Grauer 2016 provide especially capable reviews of this literature.
2. Take, for example, Schifferle 2010; Matheny 2011; Grauer 2016.
3. Peters 2007; Anonymous 2012; Dempsey 2012; Kelley and Johnson-Freese 2013; Metz 2013; see also Ricks 2013; McGivern 2014.
4. Watson 2007; Keegan 1983.
5. On the role of military education, see Schifferle 2010; Matheny 2011. For other explanations of battlefield effectiveness, see Millett et al. 1986; Biddle and Long 2004; Grauer and Horowitz 2012; Grauer 2016.
6. Lamb and Porro 2014.
7. Biddle 2006.

8. US Army Field Manual 3-0, *Operations*.
9. Biddle 2006; Galula 2006; Trinquier 2006 [1961]; Kilcullen 2009; Demarest 2011.
10. Bladgen 2016, 195, 210–11.
11. Huntington 1957; Toronto 2017.
12. Nagl 2005; UK MoD 2010; GlobalSecurity.org 2015; Losey 2015; NTC 2015.
13. UK MoD 2010, 6-15–6-17; US Army Field Manual 5-0, *Operations Process*.
14. Anonymous 2014.
15. Clausewitz 1976 [1831].
16. Pollack 2006; Talmadge 2015.
17. Grauer and Horowitz 2012; Compare Brooks 2003; Biddle and Long 2004; Ramsay 2008; Beckley 2010; Pilster and Böhmelt 2011; Brown et al. 2016.
18. For qualitative and theoretical studies, see Hazen 1872; Upton 1878; Huntington 1957; Janowitz 1971; Nordlinger 1977; Teitler 1977; Preston 1980; Harries-Jenkins 1989; Vagts 1994; Alagappa 2001. For policy documents, see US Department of Defense 2005, 2006, 2012, 3, 5–6; Joint Chiefs of Staff 2006, 23–26; Moroney et al. 2007; Johnson et al. 2009; UK Ministry of Defence 2010; White House 2010, 27.
19. On political culture and regime type, see Nordlinger 1977; Simon 1978; Perlmutter and Bennett 1980; O'Donnell et al. 1986; Stepan 1988; Agüero 1992; Nunn 1992. On security concerns, see Waltz 1979; Cohen 1985, 1997; Tilly 1992; Posen 1993; Resende-Santos 1996. Two exceptions are Pascal et al. 1979, which uses statistical measures of military quality, but only for Middle Eastern countries, and Ralston 1990, which examines military modernization cross-regionally, but only for a small number of countries.
20. Waltz 1979; Resende-Santos 1996; Gartzke 2007; Asal et al. 2015; Cohn and Toronto 2016.
21. On civil-military conflict, see Huntington 1957; Janowitz 1971; Cohen 1997; Desch 1999; Herspring 2005; Feaver 2009. On the management of violence, see Teitler 1977; Caforio 1988; Harries-Jenkins 1989; Sarkesian 1993; Vagts 1994; Alagappa 2001. On interventions in politics, see Finer 1962; Stepan 1971, 1988; Nordlinger 1977; Simon 1978; Perlmutter and Bennett 1980; Sarkesian 1981; O'Donnell et al. 1986; Agüero 1992; Nunn 1992.
22. Katzenstein 1996; Johnston 1998; Watson 2007; Toronto 2011b, 2011c, 2015.
23. For small-*n* studies of military institutions, see Lee 2006; Oktay 2010; Heiduk 2011; Libel 2013. For descriptive studies, see Harries-Jenkins 1990; Feaver 1996, 2009; Paparone et al. 2008. For cultural or institutionalist studies of military change, see Sørensen 1994; Farrell 2001; Bond 2010.
24. On democratic transitions, see Stepan 1988; O'Donnell et al. 1986; Agüero 1992; Nunn 1992. On civilian control, see Huntington 1957; Finer 1962; Nordlinger 1977; Simon 1978; Perlmutter and Bennett 1980; Welch 1987, 1992.
25. Biddle and Long 2004, 525. See also Maniruzzaman 1987; Welch 1987, 1992; Reiter and Stam 1998; Fitch 1989, 99–100; Lyons 1961; Janowitz 1971; Desch 1999; Weber 1999a, 109; 1999b.
26. On Germany and France, see Hazen 1878; Tilly 1992; Posen 1993. See also Cohen 1985; Barnett 1992; Hacker 1993; *Economist* 2014.

27. Ralston 1990; Nunn 1992.

28. Resende-Santos 1996. See also Waltz 1979.

29. See, for instance, Cohen 1985; Caforio 1988; Johnston 1998; Kier 1999; Heiduk 2011.

30. Farrell 2001, 65.

31. Lutterbeck 2013, 28; Springborg and Clement 2011; Pollack 1996; Biddle and Long 2004; Asal et al. 2015; Cohn and Toronto 2016.

32. Inglehart 1997.

33. Lerner and Pevsner 1964.

34. Nickerson 1940; Hackett 1962; Durkheim 1964; Huntington 1968; Tilly 1992; Rosen 1995, 1996; Smith 1998 [1776]; Elias 2000; Beckley 2010.

35. Pascal et al. 1979, 38; Pascal 1980; Pollack 1996, 148.

36. Caldwell and Finney 2011.

37. Beattie 2001.

38. Kollias 1996; Powell 2012; Kentor and Kick 2015.

39. Toronto 2017.

40. Watson 2007; Schifferle 2010; Matheny 2011.

41. Box-Steffensmeier and Jones 2004.

42. Toronto 2017; Hintze 1975.

43. As measured from one standard deviation below the mean of *GDP per capita (log)* to one standard deviation above the mean.

44. This first difference is based on a univariate logit estimation of untransformed *GDP per capita* on the *war college* variable. Note that this is a different model from the first differences mentioned above and reported fully in the appendix, which controls for a host of variables and uses log-transformed *GDP per capita*.

45. Schifferle 2010; Matheny 2011.

46. Bennett and Checkel 2014.

47. Showalter 1971; Posen 1993; Corum 2000; Riches 2005; Rowe 2006; Osman 2008; Pichichero 2008.

48. Pollack 1996, 2002; Lewis 2002; Biddle and Long 2004; UNDP 2016.

49. Heper and Evin 1988; Karpat 1988; Abu-Bader and Abu-Qadr 2003; Altinay 2004; Cook 2004, 2010, 2012; Duman and Tsarouhas 2006; Jenkins 2007; Cizre 2011; Cole 2012; Nassif 2013; Yildirim 2013; Sadık 2013; Cilliler 2016.

Chapter 2

France and Prussia

Military Defeat and Reform

France and Prussia are the natural starting points for an inquiry into how militaries learn, since most modern patterns of military education trace their roots there.[1] An examination of military educational reform in France and Prussia also suggests that military education plays a role in battlefield success, which is one of the core assumptions of this study. Building on this assumption, the story of military educational reform in France and Prussia also reinforces the findings of the statistical analysis above: states tend to adopt the system of military reflection according to a two-stage process. The question is, do France and Prussia conform to the expectations of adoption theory, with Radical Adoption in France and Front-loaded Adoption in Prussia?

In both France and Prussia, dire military defeats catalyzed the political will to launch reforms and a developing economy created the human and material capital necessary to underwrite reforms as the decades passed, suggesting that these two factors work in tandem to create the foundations for long-lasting change. However, in Prussia the economy developed during and after military reforms, while in France the economy developed substantially before and during reforms. This allows us to distinguish between Radical Adoption (France) and Front-loaded Adoption (Prussia). The timing of economic modernization should make initial military reforms, in response to severe battlefield defeats, deeper and more comprehensive in France than in Prussia, although adoption theory expects both states to continue gradual improvements in military education after the initial period of reform.

Comparing France and Prussia is also useful for the academic interest shown in these two cases. Scholars of military reform have mined their histories for clues on how and why militaries change the way they think.[2] I move on to examine other, under-researched cases in subsequent chapters.

PRUSSIA: FROM ROUT TO REFORM

The most significant reforms in Prussian military history occurred after 1806.[3] Prussia reformed its military institutions for decades afterward, such as Helmuth von Moltke implementing mission directives in the 1850s and Hans von Seeckt preparing the German army for blitzkrieg in the 1920s, but no reforms had as far-reaching implications as those that occurred after Napoleon routed the Prussian army at Jena and Auerstadt in October 1806.[4] In the decades after the initial period of Prussian military reform in 1806–13, economic development in Prussia gave the Prussian state access to the human and material capital necessary to make reforms last. In short, the timing of battlefield defeat and economic modernization in Prussia created a front-loaded pattern of adopting the system of military reflection.

The Scharnhorst Reforms

Gerhard von Scharnhorst fathered Prussian military reform. The son of a tenant farmer, he became one of the preeminent military thinkers of his time, even before he entered Prussian service in 1801, when he enlisted on condition he be granted the opportunity to reform the Prussian army.[5]

Scharnhorst set himself a daunting task. The Prussian army was no longer the paragon of discipline and expertise it had been fifty years earlier, under Frederick the Great. Training was infrequent and did not reflect realistic combat conditions. Soldiers received poor pay, and desertion was widespread. Officer advancement was made based on seniority and nobility, not merit, and most Prussian officers did not have a university education. Administration was chaotic, and the officer corps was rife with corruption. No formal auditing system existed, and low officer pay and pensions made "most officers [feel] justified in making a profit out of the military pay and supply system."[6] The Prussian army moreover resembled the Italian condottiere system from about a century earlier than it did an Enlightenment army. One historian likened it to a "huge antediluvian penal institution."[7]

As director of military education, Scharnhorst judged that Napoleon's battlefield success lay in mobilizing the nation for the defense of the state, so Scharnhorst and other reformers tried to "[transform] military service from a despised occupation reserved to foreign mercenaries and the 'lower orders' into the highest honor of every citizen."[8] He focused the Berlin War Academy's curriculum on training future staff officers and organized the Militärische Gesellschaft in 1802, to provide a forum for current officers to discuss war on a scientific level. The forum held 160 meetings in four years, with an average attendance over 180. When the king put him in charge of a section of the general staff in 1803, Scharnhorst instituted "staff journeys," for staff officers to make reconnaissance and planning trips to sectors under their purview. He also made advancement to the general staff contingent on passing a competitive examination.[9]

Scharnhorst and other reformers met mostly with frustration before 1806, however. Most commanders simply ignored their staff officers' advice during the 1806–07 war with France. In 1805, the army tried to reorganize along divisional lines, to improve coordination between infantry, cavalry, and artillery, but it was too little, too late. When the army took the field against Napoleon, neither officers nor men were trained to take advantage of the new organization.[10]

The battles at Jena and Auerstadt laid bare the Prussian army's shortcomings, and far-reaching reform ensued even before hostilities with the French ended in 1807. Scharnhorst became chair of the Military Reorganization Committee, which more fully instituted the divisional structure, tried to make promotions based more on merit than on seniority or nobility, established schools for training officers, subordinated all military education under a Central Office for Military Education, introduced a more significant role for the general staff, created a Ministry of War, and pushed through measures to reduce corruption in the officer corps.[11]

The post-1806 Prussian army was thus very different from the one that took the field at Jena and Auerstadt. Attaining an officer's commission and advancing thereafter relied on competitive examinations. Training on the general staff gave an officer special prestige, and commanders began to take advantage of staff planning and advisement. The new Ministry of War eliminated overlapping jurisdictions and interagency rivalries, and the reformers laid the groundwork for mobilizing the nation's manpower by eliminating nearly all exemptions to conscription and creating the Landwehr and Landsturm reserve forces, as well as officer training units.[12]

These reforms gradually established the system of military reflection in the Prussian army: stable civil-military relations, institutional reflection, and a capacity for critical analysis. With respect to critical analysis, the technical specialization of the general staff and the inclusion of talented middle-class candidates in the officer corps allowed the army to achieve a previously unknown level of expertise. The scientifically based military education system and competitive officer examinations at multiple levels specified the knowledge base necessary for command.

The reforms contributed to institutional reflection by expanding the number of military schools and intellectual fora for officers, creating time and space to learn lessons from the field. In addition, expanding conscription to all citizens created a mechanism whereby future poor battlefield performance would express itself through popular backlash,[13] especially since elites no longer gained exemptions from service.

The reforms also reinforced the stability of Prussian civil-military relations. While the noble class dominated the officer corps for most of the nineteenth century, over time the middle classes integrated into the commissioned ranks, and the basic institutions that Scharnhorst advocated—a scientific officer education system, comparatively open competition for officer commissions, and a rational military hierarchy based on a general staff—endured until Hitler came to power in the 1930s and politicized the officer corps once again.[14]

The Scharnhorst reforms were a watershed in Prussian military thinking. The later Moltke and pre-World War I Seeckt reforms were not nearly as fundamental to developing the system of reflection. Moltke's most important innovations were using mission directives and technology (like the telegraph and railroads) in staff planning, and Seeckt's were training individual soldiers to use as many weapons as possible and secretly maintaining a skeletal general staff organization, to allow for rapid army expansion when the time came. In addition, reforms from 1859–61 made attendance at a War School mandatory for obtaining an officer's commission and made merit the sole basis for receiving a commission, but the king watered these measures down under conservative pressure.[15] After Jena and Auerstadt, by contrast, Prussia created an active, peacetime general staff, completely overhauled the military education system, mobilized the manpower of all Prussians, and undertook the total reorganization of the army along divisional lines. Scharnhorst's reforms established the system of military reflection in Prussia. Later reforms merely embellished it.

Jena and Auerstadt: Catalyst of Political Will

In the words of historian William Shanahan, "[Prussian officers'] complacence received a rude shock in 1806."[16] Even though the Prussians faced Napoleon's army of 185,000 with 166,000, it was quickly apparent that they lost the day due to poor organization and leadership, not their numerical disadvantage. Split into three separate groups, the Prussians did not communicate their need for reinforcements in a timely fashion, nor did commanders demonstrate the initiative that very well could have turned the tide, especially against the main French force at Jena. Able to bring concentrated force to bear on the isolated Prussian forces, the French induced general retreat along multiple axes and, in some cases, capitulation without a shot being fired. As fortress after fortress fell in the aftermath, the royal court had to relocate from Berlin to Königsberg, in East Prussia. What began as defiance of Napoleon's continental hegemony ended as a complete rout.

Prussian society became much more enthusiastic about military reform. "It was the ultimate humiliation, but it was also the turning point, and the beginnings of a new climate in public affairs were marked by a novel and interesting manifestation, namely the appearance of parties, a thing so far unprecedented in Prussian history."[17] The German press lambasted the Prussian officer corps for shirking its duty and called for change. The provinces most burdened by the ensuing French occupation—especially East Prussia, which was the launching point for Napoleon's ill-fated invasion of Russia—were more willing to comply with the general mobilization of 1813 against French rule. Those provinces also more readily accepted the creation of the Landwehr and officer training units.[18] The defeats at Jena and Auerstadt thus marked a turning point in societal perceptions, a shift in the discourse of security.

The king himself demonstrated this radical shift. He "issued a series of decrees recommending immediate changes in the tactics, discipline, and organization of his remaining armed forces."[19] He called for the dismissal of all officers who left the field unwounded or who capitulated, for the opening of the officer corps to non-nobles, for reforms to officer education, and for new military regulations and revised Articles of War. In late 1806, shortly after the defeats at Jena and Auerstadt, he charged the Military Reorganization Committee to oversee the reform program.[20] As with Prussian society as a whole, the political will to launch reforms manifested itself in the royal court, as well.

The Jena and Auerstadt defeats were profound and unique. Prussia did not reform its military to this extent after fighting the French Revolutionary army in 1795, nor even after World War I, when some in the Prussian military establishment maintained that they could have continued the struggle.[21] In 1806 and 1807, by contrast, the Prussian military's will to fight broke, and its capacity to resist the French onslaught evaporated in a string of ignominious surrenders. Without question, the defeats at Jena and Auerstadt catalyzed Prussia's adoption of the system of military reflection.

Prussian Economic Modernization

If the defeats at Jena and Auerstadt sparked the most significant military reforms in Prussian history, the Prussian economy helped these reforms endure. As the nineteenth century progressed, Prussia acquired the wealth and human capital necessary to sustain the intellectual architecture created after 1806.

In 1806, the Prussian economy was hardly advanced. Over 80 percent of Prussian labor was involved in agricultural pursuits, and the economy was about two decades from large-scale industrialization. More than 75 percent of the 200,000 looms in the country were used only occasionally. Less than 5 percent of economic output was from industrial production, and the guilds were still quite strong. In France, by comparison, the power of the guilds had been declining since the Revolution, some seventeen years earlier.[22]

Prussian economic modernization was slow, as well. Workers' wages increased only slowly in the first half of the century, and the working day was limited to twelve hours only after 1870. Berlin, the largest Prussian city at the time, was one-sixth the size of London in 1800 and smaller than Paris had been four hundred years earlier. As a result, Prussia was consistently unable to accumulate wealth before and just after the 1806 military reforms. In the years leading up to reform, the Prussian treasury was increasingly pushed to the limit to provide resources to the military. In the last years of Frederick the Great's reign, for example, the Prussian military consumed fully four-fifths of the state budget, and Prussia had to pay a war indemnity to France from 1807 to 1813. In short, Prussia did not enjoy an abundance of wealth and human capital in 1806.[23]

Still, economic change was in the offing. In 1807, the king abolished serfdom on all crown lands, and guilds were abolished soon after. In 1818, Prussia abolished internal tariffs, and the middle class began to assert itself. Manufacturing exploded in Prussia from the 1820s on. Railroad mileage and mining, in particular, experienced a dramatic increase. In fact, mining production remained steady from 1800 to 1820, but more than quadrupled from 1821 to 1850. At one coal mine, where only ninety-seven miners were employed in 1810, 2,886 were employed in 1870. The mine produced only 1,791 tons of coal in 1847, but 793,294 tons in 1870.[24]

Economic development in Prussia manifested itself in human capital as well as wealth. Prussia was one of the first European countries to institute mass schooling, after the 1806 defeats at Jena and Auerstadt, and educational reform reflected the humanistic ideals of the Enlightenment. Prussia's taxation system was one of the most efficient in Europe, and its bureaucracy was a tool for molding Prussian identity based on ascetic Protestant ideals.[25] Prussia developed both its human capital and its wealth after 1806.

Thus, at the same time that military reforms changed the Prussian army, the Prussian economy developed into the economic juggernaut that dominated central Europe in the latter portion of the nineteenth century and twice fought the world—almost single-handedly—in the twentieth.[26] Military reforms and economic change in Prussia were mutually reinforcing, with economic modernization deepening and contributing to the institutionalization of military reforms. Adoption of the system of military reflection in Prussia was front-loaded, with an initial period of fairly widespread reforms followed by decades of steady improvement thereafter. The question for adoption theory, though, is whether this initial period of reform was deeper, quicker, and more comprehensive in France after 1870 than in Prussia after 1806.

FRANCE: FROM THE *LEVÉE EN MASSE* TO A "REDOUBTABLE INSTRUMENT OF WAR"

Compared to Prussia's Front-loaded Adoption of the system of military reflection, adoption in France was radical. Where adoption in Prussia began with a series of reforms that national leaders solidified over a period of decades, French reforms after the 1870–71 Franco-Prussian War were more all-encompassing and thorough initially than the post-Jena and Auerstadt reforms in Prussia.

The French experience shows how a modern economy influences adoption of the system of military reflection. France began significant military reforms in both the late Revolutionary and Napoleonic periods and after the Franco-Prussian War, but only the last of these actually led to adopting the reflection system. While the French economy began to modernize before the Revolution, this economic change only gained steam—and only reached an advanced stage—after 1850. This episode of economic modernization, followed by the catalyst of the Franco-Prussian War, underwrote radical adoption of the system of military reflection in France.

French Military Reforms, 1793–1805: Away from the *Levée en Masse*

Although post-Revolutionary French military reforms generated some characteristics of the system of military reflection, these traits did not last. In fact, the mobilization of the entire French nation for war after the Revolution decreased the fighting quality of the army, and in the mid-1790s France began to improve on the system of long-service regulars from the *ancien régime*. These reforms stagnated after the Napoleonic Wars. In terms of adoption theory, France experienced little battlefield motivation to adopt the system of military reflection in the first half of the nineteenth century, nor did France boast the wealth and human capital to adopt the reflection system fully even had that motivation existed.

The Revolution of 1789 had an immediate and deteriorating effect on the French army. Historian John Lynn notes, "Egalitarian ideas corroded discipline, while the turbulent confusion of the times resulted in a high rate of desertion. A rapid turnover in command, occasioned by the emigration of nearly 60 percent of all officers, struck the army still harder."[27] The selection and promotion of officers through seniority and elections had a particularly damaging effect on the expertise of the army, since it encouraged factionalism and put the oldest and most popular men in positions of command, rather than the most capable. Allowing more non-nobles into the officer corps probably gave the army access to more natural talent, but the officer corps suffered more from the mass emigration of experienced officers than it gained from opening officer ranks to the mass of the populace.[28]

In order to create his Grande Armée in 1805, Napoleon steered his men away from the egalitarian ideals of the Revolution, "[cultivating] pride in the military profession, and [arousing] greed and ambition by means of distinctions, privileges, and rewards."[29] Napoleon's use of 20–30,000-man corps, which could give battle unsupported, required officers with sufficient initiative and training to coordinate the different arms and take advantage of battlefield opportunities. Napoleon relied mainly on well-trained, long-service volunteers, not conscripts.[30]

Building Napoleon's army began even before he came to power. In November 1793, national leaders stripped elected Revolutionary officers of their ranks and incorporated their units into the regular army, then in February 1794, illiterates were barred from both commissioned and noncommissioned officer ranks. By 1795 about 80 percent of junior officers could read and write. The selection and promotion of officers began to be based more on merit and ability than on age or popularity. The quality of training improved as well—the French army evolved a coherent tactical system based on column formations, skirmishers, and concentrated artillery fire, and imparted this doctrine using experienced noncommissioned officers. Raw conscripts were gradually replaced by long-service volunteers, and permanent camps were established to facilitate large-scale maneuvers and standardize training. These reforms helped create the army that defeated Austria and Russia in 1805 and the vaunted Prussians in 1806.[31]

Nonetheless, any professional characteristics that the French army enjoyed under Napoleon gradually gave way to complacence bred of success in colonial wars. Though long-service regulars remained the norm—one of Napoleon's lasting legacies—the French officer corps never adopted a scientific approach to war, as the Prussians did during the first half of the nineteenth century. French battle tactics generally involved frontal assaults, full of élan, which served in colonial settings against poorly entrenched natives, but which proved disastrous in 1870.

The French army in 1870 was poorly organized and even more poorly led. Most telling is French officers' lack of foresight prior to the outbreak of hostilities. After the Prussians' convincing defeat of the Austrians in 1866, the French high command resisted the incorporation of new cannon and an effective reserve into the army, citing the former as too costly and the latter as too detrimental to troop quality. As a result, in 1870 the Prussians had rifled, breech-loading cannon that were far superior to the French smoothbore, muzzle-loading models in both range and accuracy, and Prussia neutralized France's initial numerical superiority by mobilizing large numbers of well-trained reservists.[32]

But the French failure went beyond cannons and reservists. It was systemic. William Fortescue notes:

> The French had not mastered the military use of railways and their mobilization was a shambles; the standard of education and training of most officers and men was well below that of their German counterparts; the German artillery and supply services were better than those of the French; and, unlike the Prussians, the French had no general staff to keep commanders in the field supplied with a regular flow of information and advice. The failure at Sedan was not that of an army, but of a whole system.[33]

Had the French been able to follow a rational mobilization plan, they very well may have been able to attack and disrupt the Germans before they mobilized fully, fighting the war on German soil and ending hostilities quickly. France, and the rest of Europe, expected this to happen, but static French defenses favored superior Prussian artillery and, after a few days, numbers.

Still, the Prussian advantage in troop strength was never tactically overwhelming due to the French chassepot, a breech-loading rifle, and mitrailleuse, a rapid-firing precursor of the machine gun. During unsupported frontal infantry assaults, Prussian casualties piled up. The French weakness lay in training on those weapons: French soldiers received little training with the chassepot rifle prior to the war, as compared to the years of experience that Prussian soldiers received on the Dreyse breech-loading rifle.[34]

French officers were intellectually unprepared for the war. A majority did not have an academy training, and the curriculum of the academies, for those officers that attended them, emphasized rote memorization, not thought. They were "abysmally ignorant."[35] They did not understand "general problems of war."[36] About half of French officers in 1870 had only the most basic education, while every Prussian officer candidate had to graduate from secondary school and pass a stringent examination before entering into one of the military academies. Saint-Cyr, the Artillery and Engineering Academy at Metz, and the Cavalry School at Saumur—the flagship institutions of French military education—were of suspect educational value.[37]

This lack of intellectual preparation showed, even before the war. Officers were promoted based more on influence than ability, and the desire for promotion and glory often led to a lack of collaboration among officers. General officers received no training, and the resultant overreliance on experience in colonial wars led to a lack of innovation and the adoption of tactics that were unsuited to European continental warfare, especially against a foe of Prussia's caliber. What training occurred was also outdated, with tactics relying on 1831 regulations, with an impractical reliance on muzzle-loading weapons.[38]

French failure to adopt the system of military reflection helps explain its battlefield ineptitude in 1870. Emperor Napoleon III's frequent and unexpected interventions in command decisions muddied civil-military relations and confused the exact role of the military in setting security policy. These interventions crippled initiative in the officer corps. The low quality of the general staff made it unable to analyze military problems critically. French Staff Corps officers often had little or no field experience, and they did not know how to organize mobilization or transport efficiently. Where French staff officers were trained to be functionaries, with poorly defined duties vis-à-vis commanding officers, Prussian staff officers were trained to be combat commanders themselves. The relative paucity of French academies, only six to the Prussians' twenty-three, prevented the French military from reflecting on the battlefield challenges of the day. One commentator of the time summed up this deficiency in the French officer corps: "Officers were brave to excess, confident in their luck, loved by their men because they were concerned with them and were regularly seen by them: these leaders had coup d'oeil, decision, in a word all the qualities acquired by experience, but they did not understand war."[39] By 1870, the French military had not adopted the system of military reflection.

France after Sedan: Toward a "Redoubtable Instrument of War"

In the twenty years following the French army's meltdown in 1870, military reforms achieved a level of military competence nearly comparable to that of the German Imperial Army. France began with military reform laws in 1872 and 1873, then continued to reform the military over the years, culminating in measures that Minister of War Freycinet pushed through in 1889 and 1890. These measures introduced a staff system akin to the Prussian general staff, improved the quality of officer education, did away with loopholes that allowed the rich and well-educated to avoid military service, and created well-trained reserves. France reformed its military institutions wholesale. In the last three decades of the nineteenth century, the French army became a "redoubtable instrument of war."[40]

The 1872 and 1873 reform measures increased the quality of French soldiers. The 1872 law abolished replacement, by which a conscript could pay a fee to avoid military service, which had overrepresented the poor and undereducated classes in the ranks and had led the French people to hate military service. The 1873 law gave draftees the option of paying for their own training and upkeep during one year's service in the army, after which they would pass into the reserve officer corps. This gave the army access to better-educated recruits, who typically assimilated training more easily, and the rest of those drafted (wealthy or not) served the full three- to five-year term. The 1873 reform law also reorganized the army into eighteen regional corps of two divisions each, which made mobilization, training, and leadership more efficient.[41]

The reform of the French Staff Corps began almost immediately after the war. In 1871, General de Miribel, who twice served as Army chief of staff, organized the staff after the Prussian model. He even had an entire bureau dedicated to studying the German army. An 1880 reform measure implemented open recruitment for staff positions, allowing the most qualified officers from any arm to stand for the staff examination. This led to a better-organized and more competent officer corps. Finally, the reforms of Minister of War Freycinet created the position of Chief of the Army General Staff, institutionalized the staff, and fomented professional standards throughout the army.[42]

The farthest-reaching reforms were in French military education. Discipline and dedication to study were imposed at Saint-Cyr, with billeting by height instead of social class and the attachment of punishment records to cadets' permanent files. The number of applicants to Saint-Cyr and the Ecole Polytechnique (which contributed officer candidates mainly to the artillery and engineers) increased. As entrance standards increased, so did the educational reputation of both schools. In addition, a new staff school, the Ecole Supérieur de Guerre, opened in 1876, replacing the Ecole d'Etat Major and "channeling the educated and hard-working toward the top of the military hierarchy."[43] The year 1874 saw the implementation of large-scale field maneuvers, with division-size opponents, giving general officers the opportunity to evaluate their tactical training for the first time.[44]

There are other indications that the French army adopted the system of military reflection rapidly after 1870. Minister of War Freycinet brought stability to French civil-military relations by reinstituting the Conseil Supérieur de la Guerre, giving its twelve officers peacetime command over the corps they would command in war and consulting them on all military matters. This helped to standardize training and doctrine throughout the army and ensured that the French approach to war was more scientific and rational than it had been in 1870. The effect was a better-educated, more reflective officer corps and, with the reformed general staff, an institutionalized capacity for critical analysis.[45] The effect evidenced itself in 1891, when large-scale maneuvers demonstrated the proficiency of French combined arms training and tactics. Of these exercises, Marshal Ferdinand Foch—one of France's World War I heroes—remarked: "For the first time since 1870, the French army had demonstrated the results of twenty years of labour, showing itself educated, well trained, equipped with excellent armaments, led by excellent cadres, in excellent spirits—in a word, a redoubtable instrument of war."[46]

Humiliation in 1870: Catalyst for Radical Adoption

The French army's performance in 1870 created intense societal pressure for military reform, such that "the ways of the conqueror, his discipline and organization, were emulated without shame."[47] Murmurs for military reform began in France after Prussia's resounding victory over Austria in 1866, but only direct military confrontation with Prussia brought about meaningful change. What is more, French society cohered around reform in the face of defeat, and French leaders referred to the military superiority of their eastern neighbor to give continued life to the reform program.[48]

French society came together as a result of the war. At the Battle of Sedan, France lost its entire second line army, not to mention the capture of Emperor Napoleon III himself. The war "began a long process of self-examination and conscious self-regeneration that would profoundly affect the lives of all French people."[49] The pall of defeat cohered French society around the notion that reform—and especially military reform—had to be borne.[50]

French leaders used the defeat to galvanize societal support for reform, especially a national myth of revenge. Primary education became free and compulsory in 1881 and 1882, in an effort to prepare the nation for revenge. In 1880, gymnastics became mandatory for all schoolchildren. In 1882, children began learning military-style drill in bataillons scolaires, and the Ligue des Patriotes was set up to encourage rifle and gymnastics clubs. In 1889, all children born in France, regardless of their parents' origin, were considered citizens, in order to make them liable for military service. In 1893, vaccinations were required for all children. Military reform received special priority: "One French institution above all had a claim on all the energy, courage and wealth of the defeated and humiliated nation: the army."[51] Without question, the humiliation of 1870 was a catalyst for military reform and, ultimately, rapid adoption of the system of military reflection.

The French Economy in 1870: Foundation for Immediate Reform

Due to the depth of the economic changes that occurred in France after 1850, one economic historian dates the end of the *ancien régime* with the period 1850–80 instead of 1789.[52] Not only did the two decades before the Franco-Prussian War see the creation of new industries, they also witnessed dramatic growth in the size of the urban population and the middle class. The French economy accumulated tremendous wealth in this period, as indicated by growth in the size of the economy.

Economic modernization in France was stagnant before 1850. Prices had been declining from 1817 to 1850, and prior to the 1850s agricultural investment had generally been in land instead of improved agricultural techniques, indicating that rural labor was still cheap and plentiful. Urban growth from 1789 to 1846 was actually slower than it had been before the Revolution.

The French economy leapt forward in the 1850s. Where agricultural workers constituted 75 percent of the active labor force in 1790–1815 and 70–73 percent around 1830, by 1896 they made up only 44.9 percent of labor. Urban growth after 1850 was impressive as well: where the urban population was 24.4 percent of the whole in 1846, it made up 31.1 percent in 1872 and 37.4 percent in 1891. In 1800, Paris was home to approximately 2 percent of the French population, but by 1900 Paris housed fully 7.1 percent of the total population.

This urban growth fueled the diversification of agriculture, which began to produce more pasture (meat, dairy) and vine (wine, vegetables, and fruit) products and fewer cereals than it had before 1850. After 1850, urban population, rent prices, and investment in building all increased. Internal migration, which tended to be temporary in the first half of the nineteenth century, became permanent migration from the 1840s, and industrial working conditions improved. Whereas laws in 1791, 1810, and 1834 had forbidden worker association and suppressed workers' rights, an 1864 law permitted strikes. Rising prices from the 1850s on also allowed employers to make concessions to workers, and the high overhead associated with training workers on specialized machines made employers less likely to fire employees. Finally, permanent migration to labor-poor areas standardized national wage levels, and living conditions in towns began to outstrip those in rural areas. The period 1850–70 was the scene of rapid modernization in French labor markets.[53]

Economic growth sped up from 1850 to 1857, and, after the 1858–59 depression, the economy continued to grow—though at a slower rate—until 1882. France paid off its five billion franc war indemnity to Prussia by 1874, and by the late 1880s, France's per capita military expenditures were among the highest in Europe. This accumulation of wealth underwrote military reforms in France after 1870. The crushing military defeat to the Prussians sparked military reforms, but a modern economy kept them alive.

France: Radical Adoption

France demonstrates the conditions for sudden, dramatic, and radical adoption of the system of military reflection. In the absence of both a modern economy and a battlefield catalyst in the Revolutionary era, France engaged in momentary military reform. In the presence of both a battlefield catalyst and a modern economy in 1870, however, France instituted military reforms that endured well beyond the immediate aftermath of the war. In fact, the French military institutions that developed after the Franco-Prussian War—especially the well-trained, dedicated officer corps and general staff—developed the tactics that were crucial to victory in World War I. Although the French military education system has been blamed for contributing to the ideology of the offensive that reigned at the beginning of World War I, Marshal Ferdinand Foch subsequently developed a system of flexible defensive tactics that allowed the allies to halt the Germans' spring offensives and turn the tide of the war.[54] What began as reforms in the late Revolutionary and early Napoleonic eras, and stagnated thereafter, resulted in radical adoption of the system of military reflection after France's crushing 1870 defeat and after she had developed an advanced economy.

The cases of France and Prussia lend support to the central argument of this study: battlefield conditions and the state of the economy interact to influence how a military learns. In Prussia, which experienced only battlefield defeat prior to reform, the reforms that led to the adoption of the system of military reflection proceeded steadily over the course of some fifty years, while in France, where both conditions existed prior to reform, adoption was radical and sudden. The Prussians reformed existing military institutions, but the French cast aside their old institutions and mimicked those of their conquerors. Had conditions been different in Prussia—for instance, had the Prussian economy already modernized substantially prior to the 1806 defeats at Jena and Auerstadt, adoption likely would have been more sudden and radical there, as it was in France in 1870. Likewise, had France experienced the kind of defeat in the Revolutionary or Napoleonic era that it did in 1870, adoption in France likely would have proceeded at the same steady pace that it did in Prussia, after an initial period of front-loaded reform.

The French and Prussian cases leave at least two questions unanswered. How does adoption theory operate outside of Europe, and what happens in cases where the battlefield catalyst is not as sudden and overwhelming as it was for France and Prussia? The next chapter examines two cases, Turkey and Egypt, to illustrate that adoption theory does apply in a non-European context.[55] Turkey experienced constant external pressure over the course of 150 years, but not catastrophic defeat like France in 1870 or Prussia in 1806. Egypt suffered a catastrophic defeat in 1967, and began reforms thereafter. As expected by adoption theory, adoption in Egypt resembles that in Prussia (front-loaded, with sudden reform after a clear defeat and continued reform as the economy modernizes), whereas adoption in Turkey fits into a third pathway, fitful adoption, which befits a country with a slowly modernizing economy and persistent external pressure instead of catastrophic defeat.

NOTES

1. Huntington 1957.
2. Nickerson 1940; Huntington 1957; Clemente 1992; Posen 1993.
3. Craig 1984. Prussian military history extends well beyond German unification in 1871, since Prussian military institutions dominated the German armed forces even after World War I. This Prussian influence on the German military did not peter out until after Hitler came to power.
4. Görlitz 1985, 40–41, 76–77, 223–25; Corum 1992; Connelly 2006, 95–116.
5. Görlitz 1985, 16.
6. Shanahan 1966, 27.
7. Görlitz 1985, 17. See also ibid., 2, and Rosinski 1939, 39–44; Shanahan 1966, 21–33, 43–44.
8. Rosinski 1939, 67.
9. Shanahan 1966, 64, 71–73; Görlitz 1985, 17, 59–60; Clemente 1992, 172–75.
10. Shanahan 1966, 68, 85–87; Görlitz 1985, 13–14, 20–22, 25.
11. Görlitz 1985, 33–36.
12. Rosinski 1939, 38, 71–72; Shanahan 1966, 124, 128–36, 141–46, 181–85, 197.
13. Horowitz and Levendusky 2011.
14. See Clemente 1992, especially xii–26.
15. Clemente 1992, xiv, 12–15.
16. Shanahan 1966, 11. See also ibid., 34; Görlitz 1985, 26–27.
17. Görlitz 1985, 30.
18. Shanahan 1966, 105, 194–96, 208.
19. Shanahan 1966, 94.
20. Shanahan 1966, 94, 103, 109.
21. Nickerson 1940.
22. Kuczynski 1945, 14; Görlitz 1985, 55; Aftalion 1990, 54; Gagliardo 1991, 138–39; Landes 1999, 243–44.
23. Kuczynski 1945, 14; Gagliardo 1991, 332; Beck 1992.
24. Kuczynski 1945, 16–17, 25, 58; Rosinski 1939, 11, 47; Henderson 1958, 15–16.
25. Schleunes 1979; Beck 1992; Gorski 1993; Kiser and Schneider 1994; Labbé 2007; Lohmann and Mayer 2007.
26. Henderson 1958, 20; 1963, 165.
27. Lynn 1996, 3. See also Connelly 2006, 4.
28. Lynn 1996, 48–49, 68–75, 88–91; Sagan 2001, 318–19.
29. Kovacs 1946, 219. See also Keegan 1993, 350–51; Connelly 2006, 74.
30. Connelly 2006, 73–74, notes that only a tenth of Napoleon's fighting force was conscripted from 1800 to 1810.
31. Görlitz 1985, 13–14; Lynn 1996, 72–77, 93–96, 216–19, 233–34.
32. Mitchell 1984, 8–16; Morrow 1993, 222–23.
33. Fortescue 2000, 3.
34. Kovacs 1946, 230; Brogan 1970, 22–30; Mitchell 1984, 16; Herrera 2004, 249.
35. Holmes 1984, 180.
36. Brogan 1970, 20.
37. Brogan 1970, 20; Porch 1981, 37–40; Holmes 1984, 55, 180–88; Clemente 1992.

38. Brogan 1970, 21; Porch 1981, 38; Holmes 1984, 56–57, 61–63, 193–98.

39. Quoted in Holmes 1984, 180. See also ibid., 11, 21–26, 59–60, 67–72, 186; Brogan 1970, 20, 22–23.

40. Porch 1981, 43. For neorealist interpretations of these reforms, see Posen 1993, 89–120; Resende-Santos 1996, 214.

41. Kovacs 1946, 221–24; Brogan 1970, 118–20; Porch 1981, 25; Mitchell 1984, 30, 46–48; Lebovics 1988, 29.

42. Porch 1981, 42; Horne 1984, 17; Mitchell 1984, 82, 86; Görlitz 1985, 97.

43. Porch 1981, 41. See also Mitchell 1984, 89.

44. Holmes 1984, 57.

45. Porch 1981, 43. On the CSG, see Mitchell 1984, 37–48, 105–11.

46. Quoted in Porch 1981, 43.

47. Horne 1984, 15. See also p. 2 and Brogan 1970, 117.

48. Mitchell 1984, 3–14, discusses the talk of reform before 1870. This talk even bore legislative fruit—the Niel reforms—but it proved hollow: the Neil reforms did not go far enough, they were underfunded, and the minister of war who took over from Niel in 1869, General Leboeuf, quietly let the reforms die. See also Kovacs 1946, 230; Brogan 1970, 17–19; Lebovics 1988, 29.

49. Tombs 1996, 46. See also pp. 33, 49–51.

50. Porch 1981, 32; Mitchell 1984, 8–15; Fortescue 2000, 5.

51. Brogan 1970, 117. See also Tombs 1996, 52–54; Anderson 1977, 12.

52. Price 1981, xii.

53. Price 1981, 66, 71, 79–84, 96–100, 168–72, 211–15, 278. See also Lebovics 1988, 11–12, 15; Tombs 1996, 23, 151; Fortescue 2000, 13.

54. Nickerson 1940; Biddle 2006.

55. Janowitz 1975.

Chapter 3

Turkey and Egypt

The Economic Foundations of Homegrown Human Capital

France and Prussia adopted the system of military reflection amid economic modernization, but Turkey and Egypt are still modernizing their economies, albeit at different rates. Both Turkey and Egypt have engaged in substantial military reforms in their history, often prompted by negative battlefield results, Egypt on the heels of the disastrous 1967 war with Israel and Turkey a number of times over the course of 150 years, usually in response to wartime defeats on the edges of the Ottoman Empire.

In the absence of advanced, modernized economies, both countries have struggled to implement the system of military reflection. At the same time, the nature of the battlefield defeats have influenced how adoption has played out. Egypt's 1967 defeat was catastrophic by comparison to the successive, though relatively less disastrous, defeats of Turkey (and the Ottoman Empire). As adoption theory predicts, Egypt has followed a front-loaded approach to adoption, like Prussia, with the 1967 defeat resulting in an abundance of political will in favor of reform, whereas Turkish military defeats constituted less of a catalyst, so adoption there has been fitful, with progress attenuated by occasional stumbles along the way. In other words, the interaction of battlefield catalysts and economic modernization affected how Egypt and Turkey adopted the system of military reflection.

TURKEY: FITFUL ADOPTION

From the late eighteenth century onward, Turkish authorities accepted intensive French, German, and even British military training, equipment support, and officer education, but Turkish military reforms met with only limited success until the Turkish economy began to modernize about halfway through the twentieth century. Until the attempted 2016 coup, the Turkish military had robust training and education programs, incorporated modern technology and doctrine, and generally showed itself to be an effective force.[1] While these military attributes are perishable, the Turkish military was still effective in its incursions into Syria in the fall of 2016.[2] Its military education system suggested a culture of reflection, and its advanced staff system, based on the NATO model, allows it to critically analyze military problems. In other words, up to 2016 Turkey had adopted essential elements of the system of military reflection.[3]

This could be changing in the aftermath of the attempted coup. Officer education, selection, and assignment have become politicized in a way they were not before, which has serious implications for the military's ability to reflect. Turkey is also in the midst of dramatic social change, with President Recep Tayyip Erdogan and the AKP consolidating state control in one party, threatening to take Turkey in a more authoritarian, less secular direction, something that elements of the Turkish military launched a coup attempt to stop.

Since then, the Erdogan government has engaged in widespread coup-proofing.[4] This throws into flux the precise role of the military in security policy making. If civilian and military leaders do not agree on the military's political role, any analysis or reflection that threatens the status quo is likely to be seen as threatening, making it more difficult for the military to fully develop solutions to security problems and, potentially, undermining battlefield performance. Based on this, it is too early to say that Turkey has fully adopted the system of military reflection, although it is farther along than most developing countries.

Despite not having fully adopted the system of military reflection, the Turkish case demonstrates the weakness of the International Security Hypothesis, which contends that states with greater international security concerns should be more likely to adopt the system of military reflection. From 1768 to 1923, Turkey fought, and lost, wars against every great power except the United States.[5] If states make war and war makes states,[6] then one would expect that, if any military were to adopt the system of military reflection, it would be the Turkish one. Moreover, from the late eighteenth century on, a parade of Western military advisers from France and Prussia, the two most advanced armies of the time, assisted with Ottoman Turkish military reforms. Despite this, Turkish military institutions continued to lack robust professional qualities until well into the twentieth century. International security concerns, in isolation, did not lead the Turkish military to adopt the system of military reflection.

Sustained economic development did. In the first half of the twentieth century, Turkey acquired high levels of wealth and human capital, two key drivers in adoption theory. The timing of these long-term historical trends holds valuable lessons for the policymakers and military practitioners ordering and conducting security assistance operations around the globe, especially in the Muslim Middle East.

Turkey's path to adopting the system of military reflection can be divided into three eras: the Ottoman era, the Young Turk era, and the Republican era. The first era began with the reforms of Sultan Selim III (begun in 1792) and continued with a number of Ottoman reform efforts throughout the nineteenth century. The second era includes the Young Turk reforms at the beginning of the twentieth century. The final era began with the abdication of the Sultan and the creation of the Turkish republic in 1923 and continues today, including the post-World War II reforms and NATO membership halfway through the twentieth century.[7]

Ottoman Era Military Reforms

The need for military reform became obvious after Turkey's defeat in the Second Russo–Turkish War (1787–92). In the war-ending Treaty of Jassy, the Sultan was forced to recognize Russian control over Georgia and the Crimea, and the backbone of the Ottoman army, the Janissary Corps, were found wanting in terms of both training and equipment. In an effort to remedy the situation, Selim III created an entirely new army using French training and weapons. The Sultan had brand new barracks built and offered new recruits exceptionally high salaries. His reforms also standardized pensions, rations, and promotion procedures throughout the army. These reforms established military regulations and banned the insidious practice of officers overreporting the number of men in their units so that they could pocket the excess salaries, a ban that new government agents enforced. Selim also opened or revived military schools to train officers, employing French instructors who used European methods.[8]

As historian Handan Akmeşe remarks, "none of these innovations proved lasting."[9] By 1806 there were still only 22,685 men and 1,590 officers in the army, and existing training facilities could not even accommodate these. After 1794, most new recruits were Anatolian peasants and tribesmen, not residents of Istanbul accustomed to discipline and restraint in close quarters. As a result, there were frequent reports of troops making off with weapons and uniforms to form robber bands that plagued Anatolia and the Balkans.[10]

When Mahmud II succeeded Selim III in 1808, he attempted to revive his predecessor's reforms, but he had to back down in the face of Janissary opposition. Mahmud finally overcame this opposition in 1826, and a flurry of reforms ensued, though they, too, did not last. Mahmud's reforms saw the rejuvenation of the officer education system, with the establishment of a military academy in 1834. Both instructors and cadets at this academy were paid high salaries, in order to lure the most qualified candidates. The year 1826 saw the establishment of a Ministry of War, and 1836 a Council of Military Affairs, to plan and implement further reforms. In 1827, Mahmud began sending Turkish officers to various European capitals to learn European military practices and then return to the Empire to train their colleagues. The Turkish government once again began instituting a formal salary system and built new training facilities across Anatolia.

However, these changes also faltered after a few years. In 1840, Prussian officer Helmuth von Moltke, retained to help reorganize the Ottoman military, gave a dim appraisal of the Turkish officer corps, complaining that their positions were attained not by competence but by social connection. The curricula of the schools never improved to the level of their European counterparts, the trickle of officer cadets to Europe never turned into a stream, and funding for training and equipment dried up after a few years. What is more, the military academy established in 1834 did not graduate its first class until 1847.[11]

Between 1839 and 1876, known as the Tanzimat era, military reform did not fare much better. During the Crimean War, the Ottoman army neglected training and the pay and administrative services did not perform well. Although the Military Academy continued to operate—one of the few lasting effects of the Mahmud-era reforms—it produced a paltry 200 officers per year, never enough in an army that numbered 750,000 by the 1890s. The vast majority of officers were promoted from the enlisted ranks and could neither read nor write, let alone understand war in an abstract way. On top of this, the general staff section of the War College did not graduate its first class until 1860, fully twelve years after it opened its doors.

The one bright spot to emerge from this era was the continued development of Turkey's military education system. In contrast to the typical Ottoman education at religious schools, the military established a separate system in 1855 to train young men to be officers from the age of ten or twelve. The best graduates went on to the general staff course of the War College, but only after serving with a line regiment for two years. Other graduates continued on to the Military Academy. This competitive military schooling system led to many careers based on merit, although the proportion of academy-trained officers did not constitute a simple majority, let alone an overwhelming one, for decades. In the 1840s, the average size of the graduating class from the War College was twenty-five officers. This average increased to more than 100 by the 1870s, and to 500 by 1900, but it was never enough to keep pace with an expanding military.[12]

In the period 1876–1908, during the reign of Sultan Abdul Hamid II, the Turkish army atrophied. The Sultan approved all officer assignments personally, which politicized the officer corps. Officers' pay was often in arrears, and enlisted men, if they were paid at all, often received overvalued government IOUs. The paranoid Abdul Hamid did not allow field maneuvers, live-fire target practice, or the distribution of artillery shells to outlying districts. He even ordered that the Mauser rifles he purchased from Germany not be distributed to the troops until six years after they had arrived. Having well-educated officers was a low priority. As late as 1894, 85 percent of officers had no formal officer education at all, and as many as one-third of all officers may still have been illiterate.

Young Turk Era Military Reforms

The 1908 Young Turk revolution ushered in another series of short-lived reforms. The Sultan's personal influence on the military ended, and the Young Turks established a Supreme Military Council in 1909 to coordinate reforms, promotions, and appointments. In July 1909, a General Inspectorate of Education and Instruction was established to improve officer education. In addition, operational planning improved under the reformed general staff, and the Young Turks instituted regular exercises and firing practice, which had not been held since 1881.

The Young Turks also purged the officer corps of its less competent members. A 1909 law purged those officers who had gained their positions by palace favoritism, and the Young Turks made efforts to purge those who had been commissioned from the ranks. All told, approximately 10,000 officers were cashiered as a result of these reforms.

These reforms also did not last. Many of the purged officers were called back to the colors after Italy's invasion of Libya in 1911. The Balkan Wars (1912–13) and World War I undermined the legitimacy of Young Turk leadership and forced the military to expand dramatically before it was ready, and training was still inadequate. The 1909 military maneuvers in Thrace had been the first in thirty years, but after holding maneuvers again in 1910, they were not held in 1911 due to budgetary constraints. In the three years before the outbreak of World War I, these reforms could hardly reverse the decades of neglect from the Abdul Hamid era. In short, as one historian has observed, "these post-1908 military reforms aimed to change much in the Ottoman army in a short time. In practice, they failed to do so."[13]

Republican-Era Military Reforms

The presidency of Kemal Ataturk (1923–38) witnessed only token efforts at reforming the Turkish military, although by the middle of the twentieth century the Turkish military was on its way to adopting the system of military reflection. While the years after World War I saw the gradual depoliticization of the officer corps, as well as a slow increase in the proportion of academy-trained officers, promotion in these years was more often based on seniority than ability.

In the mid-1940s, military training and education programs infused new life into the Turkish military's culture of reflection and capacity for critical analysis. The gradual increase in the proportion of academy-trained officers meant they finally began to reach predominance, due in large part to the gradual institutionalization of the military education system. In September 1947, the army demobilized an entire class of conscripts, reducing the size of the army from 485,000 to 350,000, and used the savings for more thorough training. When the Soviet Union demanded guaranteed access through the Dardanelles and the leasing of bases along Turkey's Black Sea coast, the United States offered Turkey $687 million worth of military aid from 1948 to 1952. Welcoming this aid, Turkey also invited American military advisers to oversee training and joint military planning, and Turkish forces received training on US equipment. In addition, Turkish officers began attending US and European military schools, and the newly created Personnel Section of the General Staff began promoting officers based more on proven ability than simply on seniority.

Even before the advent of US military aid, though, Turkey had begun devoting more resources to the military. The year 1946 saw a 31.7 percent increase in the defense budget from 1945, and the following year witnessed a further 44 percent increase. In 1950, Turkey sent a brigade to fight with UN forces in Korea. The allies regarded the Turkish contingent as a quality force, even though it suffered withering casualties. Associate membership in NATO in 1951, and full membership in 1952, were the culmination of a five-year military professionalization effort in Turkey. This effort was not ephemeral, either. Turkey has gradually increased its military expenditures per soldier in the years since. In the 1970s, there were still thousands of US personnel in Turkey, working to further train and professionalize the armed forces.[14] In a 2012 visit to Turkey's War College, I witnessed a fully developed military education system, using seminar-based, Socratic instruction methods and a theoretical approach to understanding military problems. In sum, by the middle of the twentieth century, Turkey finally succeeded in adopting the system of military reflection, something it had not achieved previously, despite 150 years of reform efforts.

Turkish Economic Development

The relatively backward economy of late Ottoman Turkey explains why Ottoman era military reforms did not endure, while economic growth in the first half of twentieth century made Turkish military professionalization possible. Consider Figure 3.1, which plots Turkish GDP per capita, a measure of national wealth, against the number of Turkish military periodicals.[15] Military periodicals suggest a commitment to the production of a military-specific body of knowledge and provide a venue to develop a culture of reflection on the military challenges of the day. After stagnant growth in the late Ottoman period and a substantial drop around World War I, significant economic growth preceded the publication of the first modern Turkish military periodicals in the 1950s. This positive correlation between wealth and military periodicals has continued into the twenty-first century.

In the Ottoman era, the first half of the nineteenth century witnessed an absolute decline in Ottoman textile manufacturing, a mainstay of the Ottoman economy. Worse still, small urban industry felt this decline sharply, and the state-run factories established earlier in the nineteenth century had to be closed. The decline in textile manufacturing was such that even the production of cotton yarn, once a staple of Ottoman manufacture, ceased entirely in Turkey in favor of raw, unspun cotton.

Government policies did not aid Turkish economic development. Tax farms were not abolished until 1839, and until the end of the nineteenth century craftsmen faced strict regulation from both the guilds and the government. Government tariffs were oppressive. In 1847 there was a 12 percent internal tariff on silk products, a tax that at other times reached as high as 50 percent.

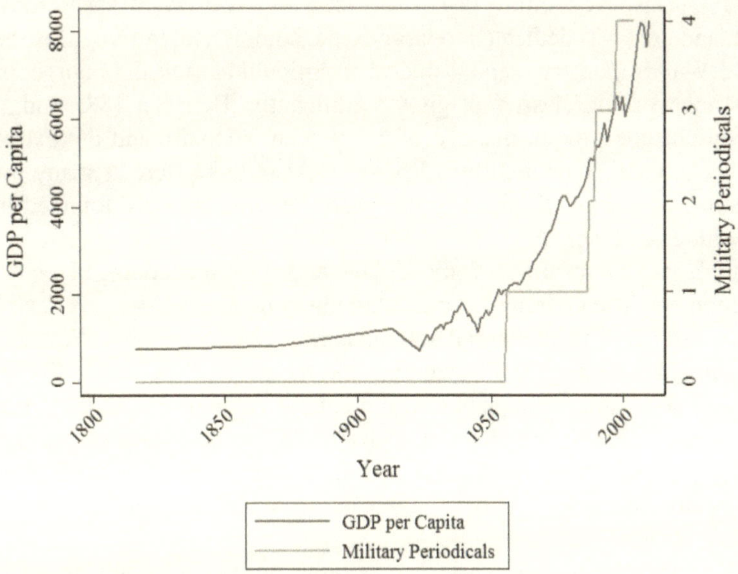

Figure 3.1 Wealth and Military Periodicals in Turkey, 1816–2011. *US Department of State.*

The requirement for joint ownership of land with the state hindered the creation of capital for new sectors.

As a result, the Turkish economy experienced little growth in the nineteenth century. In comparison to Europe, the Ottoman Empire's economy was weaker in 1800 than it had been in 1550, and may even have suffered absolute deterioration. Turkey urbanized little between 1800 and 1914, and the economy proved unable to finance the needs of the state. The government contracted nineteen loans from European creditors from 1854 to 1879, and "the greater part of what was received was used to pay the principal and interest on the debt itself."[16] In the fiscal year 1872–73, the Empire spent twice as much on servicing public debts as it did on the army and navy. Turkey's economic situation became so dire that in 1876 the Ottoman Empire declared bankruptcy.

In the absence of a modern economy, Ottoman era reforms failed to generate real change in the military. On the one hand, there was a consistent shortage of monetary capital with which to fund reforms, as illustrated by persistent Ottoman indebtedness and the Ottoman Empire's inability to pay officers and men consistently or to dedicate significant resources to the military. On the other hand, there was also constant scarcity of human capital. The military education system was never robust enough in the Ottoman era to provide the officers that the army needed, and the Ottoman army exhibited suspect levels of literacy and training.

In the Republican era, the modernization of the Turkish economy made adoption of the system of military reflection possible. Even before World War I, Turkish agriculture began to focus on a variety of cash crops and more land area was dedicated to large-scale, commercial farming. Just before World War I, industry, capital-intensive agriculture, and the bourgeois and industrial working classes all grew significantly. Between 1885 and 1913, coal output grew by an average of 8.7 percent annually and the extraction of other minerals increased by 15 percent. Also, there were as many private industrial companies founded in the years 1901–15 as were founded in the entire nineteenth century.

The Kemalist reforms of the 1920s and 1930s encouraged economic development. The government extended the railroads by three thousand kilometers, established commercial and industrial banks, adopted a Western civil code, alphabet, and system of weights and measures. Whereas in 1943 there were 421 joint stock companies with a capital value of 245 million Turkish pounds, by 1950 this number had grown to 782, with a value of 875 million Turkish pounds. This growth continued thereafter—from 1951 to 1957, 1,703 new companies were founded, representing a capital value of 1.5 billion Turkish pounds.

The growth in overall Turkish economic productivity was also significant after World War I. GDP per capita, adjusted for inflation, increased

86.6 percent from 1923 to 1938. Where urbanization had been stagnant in the nineteenth century, one-quarter of Turkish citizens were living in urban areas by 1960. There were also advances in Turkish education. The growth in the number of students in schools far outstripped population growth from 1923 to 1938; the population grew by 43.3 percent, while the number of students enrolled in school increased by 163 percent.[17]

Turkey had an increasing amount of human capital and wealth at its disposal to put toward reforming the military. It is no coincidence, then, that the resources devoted to the Turkish military increased dramatically as the Turkish economy developed. It is also not a coincidence that Turkish military training programs became more rigorous at the same time that Turkish public education and the level of specialization in the Turkish economy expanded.

Turkey: Implications for Adoption Theory

The history of Turkish military reform reveals a notable constant: considerable foreign support. Training and equipment support came principally from the French during the reign of Selim III, from the Prussians from the reign of Mahmud II through that of the Young Turks, and from the United States during the Republican era. One factor that changed from the Ottoman era to the Republican era was the level of economic development. That the Turkish military only began seriously adopting the system of military reflection at the midpoint of the twentieth century—despite battlefield defeats and reform programs for over 150 years previous—is less puzzling when one considers the main prediction of adoption theory, that adoption is most likely when conditions create the necessary political will *and* access to capital.

That said, the Turkish system of military reflection is in a state of flux. In the aftermath of the attempted 2016 coup, it is possible that the government will begin to reinforce reflective military institutions once it deems the military sufficiently loyal, but it is also possible that the government's coup-proofing behavior will continue and deepen, thus undermining the hard-earned capacity for reflection that the Turkish military has gained over the decades. Given this, Turkey will be an important test case for adoption theory in the coming years. If Turkey experiences a battlefield reverse as a result of this coup-proofing behavior, then the faster it re-adopts the system of reflection afterward, the more it would strengthen adoption theory.

At the same time, the speed with which Turkey has retreated from the system of reflection is startling, given that this process began over two hundred years ago, and in earnest some seventy years ago. It may be that this is the most recent ebb in the cycle of fitful adoption, or it may be genuine retrenchment.

Historically speaking, though, the Turkish case offers good reasons to disbelieve competing explanations for acquiring the system of military reflection. The first competing explanation is the International Security Hypothesis. If any state had an international security incentive to overcome internal resistance and adopt the system of military reflection, it was Turkey, given that it lost wars against all but one great power in a hundred-year span. Nonetheless, this international pressure did not result in adoption of the system of military reflection, despite numerous attempts to do so.

The evidence with respect to the Regime Type Hypothesis is a bit more ambiguous. Turkey began the slow process of democratization after the midpoint of the twentieth century, at about the same time that the most successful reforms began, but Turkey, up to the 2016 attempted coup, was far from a consolidated democracy, even though it had in large part adopted the system of military reflection. If the Turkish military continues to implement the system of military reflection, even in the face of creeping authoritarian consolidation in one party (the AKP), then it would undermine the Regime Type Hypothesis.[18] If, however, Turkish experience continues to reveal a correlation between growing authoritarianism and a rollback of professional military institutions, then the Regime Type Hypothesis would be supported.

The Turkish case also throws the Cultural Distance Hypothesis into question. In the early twentieth century, prior to Turkey launching the military reforms that are now leading to adoption, Ottoman Turkey was as culturally distinct from Anglophone and other Western countries as many Arab countries are today, with a reliance on religious authority structures and personal authority. In fact, the Ottoman army functioned like many Arab armies in the late twentieth century, with extensive coup-proofing behavior and a politicized officer corps. If Turkey's Erdogan government continues to reinforce elements of this cultural distance from the West, yet still adopts the system of military reflection, then it would undermine the Cultural Distance Hypothesis. If, as seems to increasingly be the case, military reflection in Turkey is in short supply at the same time that coup-proofing and politicization increase, then the Cultural Distance Hypothesis would be strengthened.

The Egyptian case brings more clarity to the Cultural Distance and Regime Type Hypotheses. Egypt was firmly undemocratic prior to the 1967 war, and it remained so afterward. Egyptian society has grown more culturally distinct from the West since, as well, with the rise of political Islam and religious conservatism in Egypt. If Egypt continues to adopt the system of military reflection under these conditions, it will suggest that neither democratization nor cultural proximity to the West is necessary for adoption (although this would not refute the notion that democratization and cultural proximity to the West can aid adoption). If the Egyptian experience also matches the expectations developed in chapter 1, with a lack of economic modernization and a severe battlefield defeat leading to front-loaded adoption, as in Prussia, then the case for adoption theory will be stronger.

EGYPT: FRONT-LOADED ADOPTION

Compared to Turkey, Egypt is not as far along in adopting the system of military reflection. As in Turkey, the exact role of the military in setting security policy is in a state of flux, with at least two very recent military interventions into domestic politics. Unlike in Turkey, however, evidence on the military education system in Egypt is sparse, and only in recent years has it become apparent that Egyptian officers who study in American military schools will be empowered to influence military decision-making, which suggests that there is some room for reflection on military problems. While Egypt has not adopted NATO doctrine and staff procedures like Turkey (or non-NATO countries like Australia), it has converted a force focused almost exclusively on external security threats to a force, since 2013, that has fought a counterinsurgency campaign in the Sinai Peninsula, which suggests some capacity for critical thinking. Military thinking in Egypt is thus not on the level of a Canada or a New Zealand, or even Turkey, but it has made consistent strides since 1967 toward adopting the system of military reflection. It is worth asking how Egypt has attained the success it has.

Egypt has engaged in significant military reforms at two times in its history: in the first half of the nineteenth century and after the 1967 war with Israel. A significant episode of economic modernization preceded the second episode, but not the first. The ambitious Egyptian leader Muhammad Ali, who ruled from 1805 to 1849, tried to reform the military on a Western model, but his reforms did not outlive him. The latter episode of Egyptian military reform illustrates how a battlefield catastrophe and a steadily modernizing economy contribute to front-loaded adoption of the system of military reflection.

Egyptian Military Reform, 1811–39: The Ambition of Muhammad Ali

The irrepressible ambition of Muhammad Ali Pasha drove Egypt's first experience with military reform. Officers and men received good training and the military enjoyed a sizeable state capital investment. In 1811, Muhammad Ali began sending a few of his officers to Europe, anticipating that they would come back to train his army at home. In 1816, he established a training course for his officers, as well as a training camp at Cairo's Citadel and a number of specialized military schools in the years thereafter. In the early 1820s, Muhammad Ali also founded a printing press, one purpose of which was to publish military manuals.

These reforms led to battlefield success. In the early 1820s, Muhammad Ali's army suppressed a rebellion in a province from which they had been conscripted themselves, routed an enemy force reportedly ten times their size in the Hijaz, and successfully protected the regime against what was rumored to be a revolt by disaffected Albanian officers.[19]

But the Egyptian armed forces hardly matched the quality of European armies. Muhammad Ali's officer corps was highly politicized. The highest ranks were reserved for his personal slaves and any promotion at all depended on both ethnicity and unquestioning loyalty to the Pasha's rule. On top of this, discipline went only so far as the officer's lash. Of those soldiers who could not avoid life-long military servitude by flight or self-mutilation (and sometimes even the maimed were conscripted), one in three deserted. The Egyptians had nothing akin to the Prussian General Staff, or even the French Staff Corps.[20]

To make matters worse, any success was short-lived. The European-imposed 1838 Commercial Convention, which limited tariffs on all imports into Egypt to 5 percent and abolished all monopolies, and the 1939 Treaty of London, which drastically reduced the size of Egypt's army, curtailed Muhammad Ali's ability to further develop his military. Before the British invasion of 1882, officers' pay was often months in arrears, and after the invasion Egyptian officers could not gain promotions without demonstrated loyalty to British rule.

The overall expertise of the Egyptian officer corps decreased considerably. Under effective British control until 1937, and then under the control of the British-leaning monarch Farouq until 1952, Egyptian officer recruitment and selection remained politicized and class-based. Only two military periodicals were published in Egypt before the rise of the Free Officers in 1952: *Jaridat Arkan Harb al-Jaysh al-Misri* was published for a short time after 1873, and *Majallat al-Jaysh* ran from 1938 to 1947. Although Muhammad Ali constructed a military with modern characteristics, including rigid discipline and passable training procedures, the reforms were neither complete nor enduring. The Egyptian military atrophied under the rule of the British and, later, King Farouq.[21]

EGYPT MILITARY REFORM AFTER 1967: A RESPONSE TO BATTLEFIELD DEFEAT

Even after the Free Officers took power in July 1952, professionalism in the Egyptian military continued to languish. Kenneth Pollack observes that by

1967 the Free Officers had "turned the Egyptian armed forces into their private fiefdom, systematically replacing all of the top military leaders with men loyal to themselves."[22] After coming to power, the Free Officers dismissed 400 of the army's highest ranking officers, replacing them with politically loyal candidates. Not only was investment in the military limited in the early years of Nasser's regime, but the Revolutionary Command Council also maintained the old class-based conscription system from the Farouq era, a system that exempted more capable, college-educated Egyptian men from military service.[23]

This state of affairs changed with Egypt's disastrous 1967 defeat and loss of the Sinai Peninsula. The army emphasized education after 1967, conscripting some 400,000 college graduates between 1971 and 1973. The length of service increased, and training improved. Some Egyptian army units practiced the Suez Canal crossing literally hundreds of times in preparation for the 1973 October War. Egyptian military expenditures per soldier increased sharply after 1967, as did the publication of military periodicals. Historian John Lynn notes this change in the Egyptian approach: "The Egyptians . . . moved away from politicized patterns as a reaction to . . . humiliation in 1967."[24]

There is reason to believe that the Egyptian military has continued to develop, albeit at a glacial pace. Since 1979, the United States has committed over $35 billion to modernize and develop the Egyptian military. Most of this has been devoted to purchasing and maintaining modern weapon systems like the F-16, which does not serve to increase the expertise of the native force, but Egypt has used over $67 million to send 1,200 officers and 4,500 enlisted personnel to US military schools. Some of these officers have returned to teach in Egyptian military schools, such as the Nasser Higher Military College.

While the Egyptian military is notoriously secretive and insular, an examination of data from the International Military Education and Training (IMET) program suggests that the stage is set for fully adopting the system of military reflection in the future. As an American program officer told me when I visited Egypt in 2010, the United States gets "much more bang for the buck" out of the $1 million or so spent on IMET programs for Egypt each year than the $1.3 billion spent on military aid.

This IMET data provides useful indicators of Egyptian progress toward the system of reflection because it contains individual-level information on programs and schools attended, lengths of study, the level of officer education (basic, junior, or senior), and the service students came from. While this data is not releasable to the public, I use it here to provide unique insights into Egyptian military development. Since Egyptian, not American, officials decide how IMET funds are spent, changes in funding reveals a great deal about military educational priorities in Egypt.

The years since 1984 have seen a forty-five-fold increase in IMET funding, in inflation-adjusted terms, including a dramatic increase in IMET funding beginning in the late 1990s. Perhaps more important, this increase coincided with a shift in emphasis toward professional military education (PME) beginning in 1998. Prior to this year, most IMET funding was spent to send officers and NCOs to short training courses in the United States. This shift is notable because, since 1979, Egypt has had to use an increasing proportion of US military aid to maintain the military hardware it has already acquired. Prior to 1998 only a fraction of Egyptian IMET students attended schools in the US PME system, but since then about one-third of Egyptian IMET students have attended PME programs as opposed to training programs. Since education tends to be more conducive to developing institutional introspection and a capacity for critical analysis than training, this is a positive sign for the future adoption of the system of military reflection in Egypt.

This shift is important for another reason. Until 1998, no Egyptian officer sent to a US PME program was from the Army, which makes up about three-fourths of the Egyptian military, in manpower terms. Since 1998, however, most of the increase in attendance at PME programs has been composed of Egyptian army students (see Figure 3.2). This shift began before the 9/11 terrorist attacks, which suggests that it was not originally connected to concerns over terrorism. By 2010, the proportion of IMET students from the Egyptian army was commensurate with its size relative to the Egyptian Navy and Air Force.[25]

This shift to PME has increased the time that Egyptian officers spend in American military schools. Until the year 2000, the average duration of PME programs for Egyptian IMET officers hovered around two or three weeks. Since then, this average has been four-and-a-half weeks or longer. This increase has come despite huge increases in the number of Egyptian officers attending basic- and junior-level PME courses, which tend to be of much shorter duration than senior-level courses, which typically last at least forty-two weeks and take place at staff and war colleges. In other words,

Figure 3.2 Egyptian IMET Students, by Service, 1978–2010. *US Department of State.*

enough Egyptian officers have attended senior-level PME courses (staff and war colleges) since 2002 that the average course duration has increased.

IMET—which is separate from the commonly cited $1.3 billion in US military aid to Egypt—is not the only source of funding for Egyptian professional officer development in the United States. While the increase in the number of Egyptian army officers studying in the United States in the late 1990s is dramatic, Egypt has also used Foreign Military Financing (FMF, part of the oft-cited $1.3 billion) to send officers to the United States. So, the number of Egyptian officers that have studied in US military schools is even greater than that depicted in Figure 3.2. In fact, since 2000 three Egyptian officers have attended the US Army's prestigious School of Advanced Military Studies (SAMS), which involves passing a competitive entrance examination. This is something no military officers from other Arab militaries have achieved, though they have tried. Furthermore, Egypt likely sent these officers to SAMS using FMF instead of IMET funds, as the IMET data do not contain records of their attendance. This illustrates the Egyptian military's commitment to improving the human capital of key up-and-coming officers.

The Egyptian cohort in US military schools is drawn from across the officer corps, as Figure 3.3 shows. This figure depicts attendance at US military schools, divided by the three main levels of officer education: basic (initial post-commissioning courses), junior (company-level qualification courses), and senior (staff and war colleges). While the increase in Egyptian IMET students beginning in the late 1990s was due mostly to attendance at basic and junior courses, there was still an increase, from the early 2000s on, in the number of students attending staff and war colleges in the United States. This is depicted in Figure 3.4, which breaks down these senior-level attendees by service. Again, while it is noteworthy that the number of Egyptian army IMET students experienced an increase in the early 2000s, it should be remembered that many Egyptian officers attended US schools using FMF funds. Their development in US schools is not captured in these data.

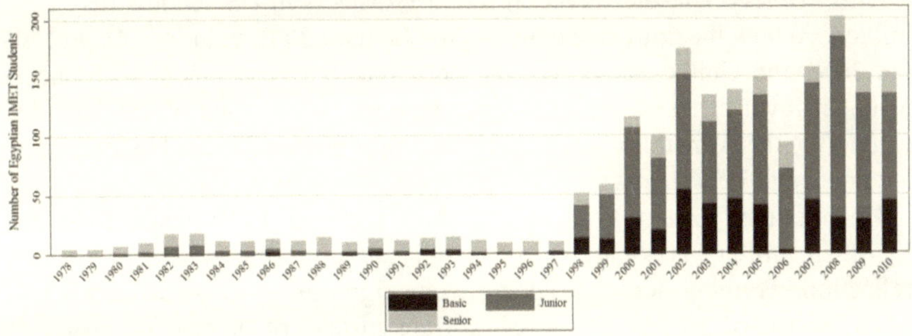

Figure 3.3 Egyptian IMET Students, by Education Level, 1978–2010. *US Department of State.*

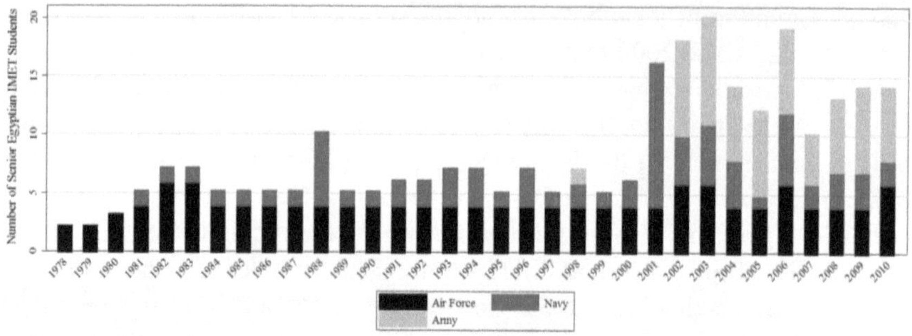

Figure 3.4 Egyptian IMET Students at Senior-Level US Schools, 1978–2010. *US Department of State.*

While there is some evidence that Egyptian graduates of US schools were kept out of influential positions in the Egyptian military before the January 25 Revolution, recent events indicate that having attended a US military school may not be as much of a career detriment as once thought. General Abd al-Fattah al-Sisi and Major General Sedky Sobhy, appointed on August 12, 2012, as minister of defense and Egyptian Armed Forces Chief of Staff, respectively, are graduates of the US Army War College, a senior-level course in Carlisle, Pennsylvania.[26] General Sisi's rise, furthermore, could be described as above-average; in the six years since being at the War College, he rose three ranks, from brigadier general to general, and was appointed minister of defense. After his retirement from active service in 2013, he became president of Egypt.

Does this mean that attendance at a senior US military school contributed to General Sisi's advancement? In a country that is hypersensitive to the influence of foreigners, this is almost certainly not the case; it is much more likely that he was simply a very competent military officer. Still, General Sisi's having attended an US military school was not a significant roadblock in his military career, which suggests an increasing value for critical thinking and analytical ability in the Egyptian military.

This effort to educate Egyptian military officers in US schools may have influenced how the army responded to the January 25 Revolution. As protests unfolded, the United States was the only Western government with access to upper-level officials in Egypt, and the conciliatory tone of the Supreme Council of the Armed Forces (SCAF) may have been influenced by members' attendance at US schools.[27]

Despite the positive role of attendance at US PME institutions, the Egyptian military is some way from fully adopting the system of military reflection. Training does not appear to be rigorous or frequent, and before 2013 it focused on force-on-force warfare instead of on the full spectrum of operations. Officer advancement still seems to depend more on personal connections than merit. Egypt coproduces the M1A1 Abrams tank and (until 2009) hosted the United States and other countries in biannual Bright Star training exercises,[28] but there is little standardization of equipment and processes in the Egyptian military and most maintenance is performed by contractors, not uniformed personnel. Egypt acquires an impressive amount of hardware with US military aid, but there is little confidence that the Egyptian military could use this hardware to its fullest extent. So, while the humiliation of Egypt's armed forces in 1967 spurred the government to launch reforms geared toward adoption of the system of military reflection, actual adoption since then has been modest and slow.

ECONOMIC MODERNIZATION IN EGYPT

Little economic modernization occurred in Egypt in the two centuries preceding the 1967 war. Investment in industry was rare in nineteenth-century Egypt. At the time of the French invasion in 1798, most manufacturing in Cairo was on a small scale and confined to the narrow alleys and dead end warrens of the old city. Cairo boasted more than a quarter-million inhabitants at the turn of the nineteenth century, but the industrial sector only employed a few hundred workers. Capital was mostly tied up in tax farms, not in ships, shops, or other productive enterprises.[29]

Muhammad Ali launched an industrialization program, but he could only man and finance it by draining the countryside of its inhabitants, by instituting a system of monopolies that extended even to basic foodstuffs and cash crops, and by increasing the land tax to the limit of peasants' ability to pay. Even with this tremendous effort, at its peak the number of industrial workers—30,000 in 1838—constituted just over 1 percent of the total Egyptian population. The factories never ran at full capacity, and in the 1830s Egypt had less than ten steam engines. Due to a lack of managerial and entrepreneurial skill, coal, iron ore, and engineers, the output of Muhammad Ali's factories declined over time, rather than improved. Muhammad Ali's monopolies program sometimes sold goods below cost, whether to get rid of them or because accountants simply did not know how much units actually cost.

Even without the 1838 Commercial Convention and the 1839 Treaty of London, Muhammad Ali's industrialization program likely would have died anyway. Machines, including the steam engines, were dilapidated from years of neglect, and sickness among Egypt's cattle posed a serious problem of power.[30]

The productivity of Egypt's nineteenth-century economy was poor. An 1836–37 decline in world cotton prices sent Egypt's economy tumbling, evidence that the economy relied too much on cotton. This malaise did not improve in the last half of the century: Egypt contracted eight international loans from 1862 to 1873, and in 1876 declared bankruptcy. This bankruptcy caused the failure of an industrialization scheme that state leaders had instituted to reverse the ill effects of Muhammad Ali's experiment. All told, throughout the nineteenth-century Egypt's economy became more dependent on cotton exports, not less, and the gains from these exports, which were considerable during the American Civil War, were not reinvested in developing other sectors of the economy.[31]

Egyptian urbanization remained stagnant until the 1920s. Urban dwellers made up 10 percent of the population in 1800, 12.8 percent in 1882, and only 15.6 percent in 1917. This slow urbanization, coupled with industrial inefficiency, poor economic performance, and the lack of sectoral diversification, indicates that the Egyptian economy was not significantly modernized by the time Muhammad Ali began reforming his armies.[32] Muhammad Ali's military reform effort also had significant external help: French officers led the program in-country and Muhammad Ali sent Egyptian officers to France for education and training. In spite of this support, Muhammad Ali's military reform project still failed to achieve its goals.

Around the middle of the twentieth century, the Egyptian economy began to modernize in earnest. Even during the period of Arab socialism from 1956 to 1967, Egypt's labor market experienced substantial sectoral dispersion. Egypt's economy was heavily dependent on agriculture (especially cotton exports) at the beginning of Nasser's rule, but from 1955 to 1967 agricultural production as a percentage of national output decreased from 32.3 to 27.5 percent, while industrial production increased from 17.6 to 28 percent of national output. From 1965 to 1980, Egypt's industrial sector grew by 7 percent annually and the service sector by 9.5 percent, while the agricultural sector grew by only 2.8 percent. The proportion of Egyptian workers employed in agriculture decreased from 60 to 54 percent between 1950 and 1970, while in the same period the proportion of industrial workers increased from 12 to 19 percent.[33]

Government policies after 1952 also favored economic modernization. Although Nasser nationalized many foreign capital interests in 1956 and domestic ones in 1961, he began reversing this trend after 1967. Nasser also reduced taxes for domestic and foreign capital interests and reformed land, monetary, fiscal, and tax policies, further fomenting growth. As a result, the business and professional classes that emerged in the 1920s and 1930s grew into a dominant force by the 1950s. Manufacturing plants employing ten or more persons grew from 95,000 in 1927 to 265,000 in 1954, an increase of over 180 percent. Industrial production increased at an even faster rate in the same time span.[34]

The timing of Egyptian economic modernization helps explain why military reforms ultimately failed under Muhammad Ali but seem to be progressing, albeit slowly, under Nasser and his successors. In both eras, Egypt benefited from Western expertise and equipment: from France and the United Kingdom in the nineteenth and early twentieth centuries and from the Soviet Union and the United States more recently. The extent of foreign training efforts does not seem to have been pivotal for successful Egyptian military reform.

The one significant difference between these two eras is that, in the nineteenth century, Egyptian military reformers did not benefit from the wealth and human capital that a modern economy provides, but after 1967 they did. In Egypt, as in Prussia, the long process of economic modernization began just prior to a long-term military reform project, which substantiates the expectation of adoption theory that, while an economy is modernizing, a catastrophic military defeat like the 1967 war with Israel will result in front-loaded adoption.

ADOPTION THEORY AND ACCESS TO CAPITAL

Like the Turkish case historically, the Egyptian experience aligns with the expectations of adoption theory. Adoption in Egypt has been front-loaded, beginning with moderately dramatic reforms after its 1967 defeat, and progressing slowly thereafter, but only in the presence of a gradually modernizing economy.

Despite their general agreement with adoption theory, the evidence presented thus far does not directly link a modern economy and adoption of the system of military reflection. The four cases of France, Prussia, Egypt, and Turkey establish a convincing correlation, but not necessarily causation. Thus far, the evidence indicates a two-stage interaction between international security concerns and economic modernization in shaping pathways toward adoption, but it is unclear precisely how economic modernization institutionalizes adoption of the system of military reflection.

The evidence presented thus far is nonetheless suggestive. The statistical evidence shows a correlation between wealth accumulation and the system of military reflection from 1816 to 2011, while controlling for competing explanations, and the case studies describe dramatic economic growth prior to effective military reform. The statistical analysis also points to a clear association between human capital and the presence of war colleges, at least since 1970. Likewise, in Prussia, France, Turkey, and Egypt, increased schooling and literacy rates and the consistent development of military education systems suggest a link between human capital and the presence of the system of military reflection.

How do human capital and wealth actually change the way a military thinks? Traditional research methods are less helpful for answering this question, especially for cases like Egypt and Turkey. Scholars of historical Latin

American military professionalism have relied on interviews with local military officers or analysis of military periodicals to better understand the transition from military rule to democracy,[35] but military periodicals are generally not available in the Middle East, and establishing the necessary contacts to conduct interviews would take years of patient engagement.[36] Interview subjects still might not be able or willing to speak freely, given surveillance from intelligence services or negative perceptions of foreigners. The archival approach has also not proved promising. When I searched Egypt's National Library in 2010 to find materials on the US-Egypt security relationship, I was met with bureaucratic foot-dragging by low-level functionaries and probing questions about my background and intent.[37] Such traditional approaches might also expose researchers to detention by security services, particularly in uncertain security environments, as those in Egypt and Turkey.

Even officially sanctioned interviews are likely to be of limited utility. For example, the Egyptian government has denied requests for US officials to meet on an informal basis with Egyptian alumni of US military schools or officers who have attended military-to-military engagement seminars and workshops run by the US government, particularly since the fallout from the Egyptian transition of power in 2013. Even before the 2013 coup, contact between the Egyptian and US militaries tended to be with only a handful of Egyptian officers, and usually on a predictable pattern (say, weekly or monthly meetings). Only under a different set of circumstances would it be feasible to collect direct evidence of the link between human capital and wealth, on the one hand, and adoption of the system of military reflection in Egypt and Turkey, on the other.

The next chapter presents direct evidence of this link, but for a different country: the United Arab Emirates. In it, I reflect on five years of experience at the UAE's National Defense College, where I worked closely with senior Emirati military officers and civilian officials in an effort to help them adopt key elements of the system of military reflection: institutional introspection and a capacity for critical analysis. Thus, the story moves from the broad-brush correlation of the statistical analysis, through four historical cases that substantiate the main expectations of adoption theory, to a quasi-anthropological field study using process-tracing methods.

NOTES

1. Birand 1991; Altinay 2004; Jenkins 2007; Yinanç 2016.
2. Stein 2016.
3. Altinay 2004; Erkuş 2016.
4. Talmadge 2015; Yinanç 2016. See also Heper and Ahmet 1988; Karpat 1988; Jenkins 2007; Cizre 2011; Çelik and Gür 2013; Cilliler 2016.

5. Issawi 1980, 4. The Republican era is typically considered to have begun in 1923, when Kemal Ataturk founded the modern Turkish republic. See also Karsh and Karsh 1999.

6. Tilly 1992.

7. Lewis 1968, 39, 56–57; Hurewitz 1969, 35–39; Shaw 1971, 112–37; Hale 1994, 12–14; Goldman 2002; Ahmad 2003, 25–29; Akmeşe 2005, 2–7.

8. Berkes 1965, 55–60, 71–72, 75; Lewis 1968, 49–50, 56; Shaw 1976, 257–60; 1971, 114–16, 119, 130–31; Ralston 1990, 47; Hale 1994, 14–16.

9. Akmeşe 2005, 3.

10. Shaw 1971, 132–33.

11. Berkes 1965, 82–85, 89–91, 102, 111–12; Lewis 1968, 70, 78–84; Be'eri 1969, 301–2; Hurewitz 1969, 36–37; Owen 1981, 58; Ralston 1990, 55–57, 62–63; Hale 1994, 16–21; Akmeşe 2005, 4–6.

12. Be'eri 1969, 303; Hurewitz 1969, 38; Ralston 1990, 59–62, 64–65; Hale 1994, 23–24, 29; Akmeşe 2005, 2, 6–8.

13. Akmeşe 2005, 120. See also ibid., 109, 115–21; Ralston 1990, 70–71; Hale 1994, 29, 42–46, who notes that, by World War I, "[Turkey's] fighting forces remained woefully defective in training and equipment" (46).

14. Pitman 1988, 318; McGhee 1990, 42–44, 77–78, 92; Criss 1993, 331–44; Hale 1994, 96–97; Smith 2000, 68; Ahmad 2003, 104–7.

15. National income figures are from the Maddison Project 2016, and are in constant (1990) US dollars. Data on military periodicals come from FirstSearch 2005; Ulrich's Periodicals Directory 2005.

16. Owen 2003, 105. See also ibid., 58–59, 62, 100–105, 118–19, 122; Aktan 1966, 111; Issawi 1966, 46; 1979, 103–10; 1980, 1, 11, 33–34; Ubicini 1966, 43; Sarç 1966, 48–51, 54–57.

17. Berkes 1965, 134; Novichev 1966, 68; Issawi 1979, 111–12; 1980, 13–15, 37, 202, 204, 272–74, 277, 367–70; Richards and Waterbury 1990, 187; Hale 1994, 52.

18. Duman and Tsarouhas 2006; Jenkins 2007; Cizre 2011; Çelik and Gür 2013.

19. Hurewitz 1969, 30; Ralston 1990, 87, 91–92, 94; Fahmy 1999, 64–66.

20. Ralston 1990, 91; Fahmy 1999, 64–72.

21. Issawi 1966, 361–63; Hurewitz 1969, 34; Barnett 1992, 60–61, 89; FirstSearch 2005; Ulrich's Periodicals Directory 2005.

22. Pollack 2002, 58.

23. Be'eri 1969, 106; Barnett 1992, 92, 130.

24. Lynn 2003, 313. See also ibid., Chapter 8; Barnett 1992, 124–28, 219; Pollack 1996, 266.

25. IISS 2012.

26. Al-Bahi et al. 2012.

27. Albrecht and Bishara 2011; Barany 2011; Cropsey and Milikh 2011; Azzam 2012. See Dreazen 2011 for an alternative view.

28. Bright Star exercises began again in 2017. See Battles 2017.

29. Issawi 1966, 373, 452–53; Owen 1981, 47–50.

30. Hurewitz 1969, 31–32; Owen 1981, 65–72, 74–76; Fahmy 1999, 65.

31. Issawi 1966, 365; Owen 1981, 73, 122, 126–27, 150.

32. Issawi 1979, 102, 109, 111–12; Owen 1981, 24, 217.

33. Issawi 1966, 452–53; 1982, 231; Richards and Waterbury 1990, 69, 71, 74, 152; Barnett 1992, 85–86, 97 (note 65).

34. Issawi 1966, 366, 369; Barnett 1992, 83, 88.

35. Stepan 1971, 1988; Nunn 1983, 1992.

36. Campbell 2009; Springborg 2011a, 2011b; Sayigh 2012.

37. Given the secrecy and insularity of Middle Eastern militaries, open source research is also unlikely to reveal a direct link. See Lahlou 2011; Middle East Channel 2011; Elgindy 2012; McGrath 2012.

Chapter 4

The United Arab Emirates

Importing Human Capital

In late 2012, the small Arabian Gulf nation of the United Arab Emirates began an ambitious experiment. The goal was to build a National Defense College (NDC) that would be the premier school for senior level defense and security studies in the Gulf, possibly even the home of a defense college for the Gulf Cooperation Council (GCC).[1] The ultimate aspiration was to matriculate students from the United States or other Western countries. As of its fifth year of operation, NDC has not yet achieved these lofty aspirations, as all of its students so far have been Emirati nationals, but setting such a high bar at the outset is a firm statement that the country wants to step, firmly, into the modern age of military education.

Establishing the UAE National Defense College was part of a larger effort to improve the state's military capabilities and claim a place as a shaper of regional affairs.[2] Since the United States' so-called pivot to Asia and the Arab Spring protests that rocked the Middle East,[3] the UAE has sought to use its influence more directly to blunt the power of perceived extremist groups, like the Muslim Brotherhood, and Iran.

This pressure to take on a more active regional role has been building for some time.[4] Three events played an important role in this shift. First, Iraq's invasion of Kuwait in 1990 belied both Arab and Western guarantees of security to the Gulf states.[5] UAE leaders, in particular, mobilized significant military resources, for such a small country, to answer Saddam's threat, and the 1990s saw a gradual increase in the proportion of nationals serving in the UAE Armed Forces. Second, UAE leaders took a more active response to the 9/11 terrorist attacks in the United States than other Gulf leaders, openly condemning the Taliban for giving Osama bin Laden sanctuary. Thus began a pattern of UAE official intolerance for extremism, in all its forms, religious and otherwise. The third event that catalyzed the UAE's motivation to modernize its military, and use it as a tool of regional influence, was the so-called Arab Spring. These uprisings created a series of power vacuums across the region and demonstrated the need for active UAE leadership in providing regional security. Given America's earlier "pivot to Asia," the time was ripe by 2011 for the UAE to assert itself regionally in a way that it had not before.

As part of this shift to address external security pressure in the Gulf, the UAE for the first time conducted offensive military operations outside its borders without the support of Western allies.[6] In 2011, in conjunction with the Egyptian military, UAE air forces attacked targets in Libya associated with extremist militias. This event was notable in that it was conducted with neither the foreknowledge nor the assent of the United States. The UAE gained a reputation among Western military officers for operating with allies in Bosnia, Somalia, and Afghanistan in the 1990s and 2000s, but since 2011 the UAE has cooperated with Egypt, Saudi Arabia, Jordan, and Sudan, among others, to conduct operations in Bahrain, Libya, Syria, and Yemen.

Even before the UAE began to flex its muscle in this new, independent way, its military was regarded as one of the most competent in the Middle East.[7] For a number of years, its special forces contingent has been led by a well-respected former Australian officer, and a few former officers from other Western militaries have served in the UAE Presidential Guard, and possibly in the air force.[8] The UAE military engages in a regular training regimen, and the military has schools for every level of an officer's intellectual development: pre-commissioning academies, basic schools for initial and captain's level qualification, a staff college, and (in NDC) a war college.[9] Programs like National Service, in which all males between eighteen and thirty must serve in the military for twelve months, seek to build human capital in society as a whole, and identify new sources of human capital to recruit into the standing military.[10] Officials aver that combat roles in Yemen operations are reserved for Emiratis. In a country of only one million citizens, efforts like these suggest a serious commitment to improving the native quality of the force.

The NDC sits at the pinnacle of these efforts. Aides consistently brief senior national leaders on the progress of the College, and US Embassy officials also brief the American Ambassador, suggesting that this joint US–UAE project is of great importance to both governments. Although there are only a few hundred individuals in the country with the English skills and the requisite experience to attend the College, the UAE government still excuses over thirty of its most capable officials to attend the course every year. Alumni are highly placed in the government after graduating, as well. At least seven alumni serve on the UAE's National Security Council staff, and other alumni have played prominent roles in the Yemen conflict. While it is possible to overstate the importance of NDC's role in developing the system of military reflection in the UAE—two civilian universities in the country also offer Master's programs in strategy and security studies, and decision-making at the highest levels of government relies more on leaders themselves than on advisors—early indications are that NDC punches above its weight in the government. In a country as small as the UAE, even a handful of alumni in key positions suggest an outsized influence.

The core mission of the College is to develop the strategic thinking skills of future Emirati leaders, which ultimately aims to improve the country's ability to organize its resources and power in order to achieve strategic objectives and preserve national interests.[11] The College is intended as a haven to cultivate reflection and generate critical insights, and it puts senior civilian officials and military officers together for a year, so it acts as a stabilizing force for Emirati civil-military relations, as students at the College share ideas about national security from across the civil-military divide, a divide that in the Emirates has historically been wide, with few civilians having a clear understanding of the armed forces' role in decision-making and strategy. This is largely because the UAE has had a professional force from the founding of the state in 1971, and because up until the 2000s service in the military was not uniformly distributed across families and tribes.[12] Insofar as the College brings elites from across Emirati society together, it will help Emiratis to identify and reinforce acceptable civilian and military roles in national security, analyze policy options with a critical eye, and develop institutional introspection, all key elements of the system of military reflection.

Classes at the College began in the fall of 2013, and policies and procedures established thus far indicate that the UAE is pursuing a markedly different strategy of military educational reform than that of Egypt and Turkey. Whereas Egypt and Turkey have focused on slowly growing human capital at home, the UAE has imported Western faculty and technical experts to develop the College much more rapidly—and with a lower level of institutionalization—than Egypt has its Nasser Higher Military College and Turkey has its military education system. The risk that existing state institutions will push back against a potentially transformational agency such as NDC is higher than with the slower Egyptian and Turkish approaches, but the reputational rewards of a quick, early breakout in military educational reform could be substantial for the UAE.[13]

THE UAE: ADOPTING THE SYSTEM OF MILITARY REFLECTION

So far, the NDC enterprise bears some hallmarks of success, but the jury is still out on whether NDC can achieve its world-class aspirations.

Although it is still early, NDC has yet to show that the system of military reflection, as defined in this study, is firmly entrenched in the UAE defense establishment. That said, most Emirati officials familiar with this military reform project would recognize that a great deal of work remains to be done, which is why the College was established in the first place.

As for the specific elements of the system of military reflection, there is clear evidence of stable civil-military relations in the wider Emirati military, even if this is based more on personal loyalty than in Western militaries. This was especially evident after fifty-two Emirati soldiers were killed in a missile strike near Ma'rib, Yemen, in 2015. UAE leaders spent a great deal of time sitting with families of those killed and explaining what motivated operations in Yemen.[14] An event that could have turned into a civil-military crisis over the lack of clarity in the reasons for the war instead became a deeper reservoir of trust by Emiratis in their country's leadership.

In contrast to the stable civil-military relations, by Western standards of military education, the level of critical analysis and institutional introspection at NDC, and in the UAE military as a whole, is generally weak. This manifests itself in the wider military organization through an overreliance on imported contractors to run the organization, sometimes burdensome bureaucratic processes, a lack of initiative at lower levels, and occasionally unrealistic training scenarios. For example, in the past there has been a reluctance to train at night, although with the intensity of operations in Yemen and, for a time, Jordan and Syria (in the coalition against ISIS), night training has become more regular. The UAE Armed Forces also regularly train with Western, particularly American, allies. So, while the Emirati military has integrated some of the most advanced equipment in the world into its force, this native human capital deficit is something that the Emirati military is working to rectify, which indicates that the force is still in the process of adopting the system of military reflection.

At the same time, there is evidence that national decision makers are empowering NDC graduates to develop institutional introspection and a capacity for critical analysis in the country by placing them in key government positions after they attend the College, which in turn encourages more of the nation's best and brightest to apply for the course. If this pattern continues, and if NDC can qualify for accreditation in the United States, then the UAE stands a good chance of boasting the premier senior level defense college in the region, an institution that would be the crown jewel in the country's military education system. Realizing these two big "ifs," US accreditation and student placement, could be the difference between success and failure in establishing a military education system that leads to sustained regional influence in the long run.

ANSWERING THEORETICAL QUESTIONS

I am hardly an impassive observer of the NDC story. The evidence I present here is based on extensive conversations with Emirati and Western officials over the course of five years, as well as personal observation. As one of the founding faculty members at NDC, I have seen the College grow from its first year, when few expected us to produce the quality of program and graduates that we did, to its fifth year, when the challenge is to continue to exceed the expectations set. Since I have a personal stake in NDC's success, there is a chance of bias in my account, so the reader should be forewarned. But I base the credibility of my account on the knowledge that no Westerner can stay in the UAE forever, so I have an obligation to myself to maintain some level of detachment. Perhaps more importantly, by being a participant observer in the NDC story, I can give a much finer level of granularity on the processes involved, and I can comment with some authority on how students' attendance at NDC has influenced the acquisition of human capital on an individual level.

In conducting this research, I placed the highest priority on protecting human subjects. In accordance with the American Anthropological Association's "Statement on Ethics," my first obligation was to "do no harm."[15] This research is intended to improve an ongoing process of military reform in the UAE by taking an objective yet critical look at the history of NDC, and the conclusions stand to benefit both the College and the UAE. At the same time, in a state with a persistent surveillance capability, it is possible that, if this research is regarded negatively, those with whom I spoke about this subject could experience negative repercussions.[16] Given this, I have kept neither notes of interviews nor a database of those interviewed, even though I spoke with dozens of individuals over a five-year period. I rely instead on immersion in a wealth of conversations over time to arrive at understanding.

I have been open and honest about my work, as I sought and received command approval to conduct interviews to write this history of the College, and I have made this chapter and other portions of this project available to those who shared their views with me, even though doing so increases the risk of my work coming under the censor's knife. At no time have I tried to curry personal favor in order to receive privileged or classified information, but instead I have offered constructive criticism and advice about how the College can best achieve its stated aims. In doing so, I have made clear that I am an outsider, though one with an intense desire to see the UAE's goal of adopting the system of military reflection succeed.

On balance, then, my research approach protects the individuals who participated while offering conclusions in a way that serves the UAE's military reform efforts. In so doing, I have tried to be both ethical and transparent in answering this study's theoretical questions.

In that regard, the growth of human capital is a critical piece in the UAE's adoption of the system of military reflection. As described earlier in this study, adoption theory predicts whether a country that does not experience a catastrophic battlefield defeat, but that enjoys a modern economy, is likely to follow a steady pattern for adopting the system of military reflection, as opposed to front-loaded (Egypt, Prussia), radical (France), or fitful (Turkey) patterns. So, is the UAE's economy modern or not? If the UAE economy is modern, then adoption theory would expect it to undertake steady progress toward acquiring the system of military reflection. If it is not modern, then it should conform to the more fitful Turkish pattern of development.

In terms of the accumulation of wealth, the UAE's economy is clearly modern, but it is still modernizing in terms of diversification and human capital. The UAE enjoys one of the world's highest rates of per capita income, and most of the population lives in municipalities with over 100,000 people. Internet usage and technology penetration are very high. The UAE economy is diversifying away from oil—only 40 percent of gross domestic product came from the oil sector in 2014—but the economy still relies on oil in an indirect way.[17] Since oil prices dropped dramatically in 2014, the government has cut spending and reduced subsidies on commodities such as fuel. Even though consumer prices have continued to rise, especially in Abu Dhabi, a number of businesses have struggled to remain open, with hundreds shutting their doors in Dubai in one month in early 2016.

Perhaps the biggest challenge in modernizing the economy, however, is in improving indigenous human capital. Human capital flight is an endemic problem, with expatriates staying, on average, only three and a half years. The government sector, mostly staffed by Emiratis, appears to be near saturation in terms of the number of jobs it can sustain, especially given the reduced oil prices. Emiratis make up only 2 percent of the private sector workforce.[18] Emiratization programs, meant to replace expatriates with Emiratis in the private sector, have been difficult to sustain because expatriate workers are often more capable and experienced in technical and management roles than their Emirati peers. This lack of Emirati human capital extends well beyond government officials, as well. There are very few public libraries, and language barriers, the lack of a customer service culture, and poor business management practices increase transaction costs in the economy and make it difficult to accomplish simple, everyday tasks.

This imbalance between wealth and human capital makes the UAE an intriguing case to study, because it isolates the role of wealth in adopting the system of military reflection. If the reflection system can simply be bought, and human capital in the wider economy is unnecessary for adoption, then the

UAE case should show evidence of ready adoption of the system of military reflection. If, however, human capital in the wider economy is necessary for adoption, then the UAE case should show adoption of the system of military reflection only as human capital improves.

The UAE is thus a good case with which adoption theory can be tested. In the language of process tracing, a social science case study methodology, the historical accident of capital imbalance in the UAE economy sets up a hoop test to examine the process connecting human capital and adoption of the system of military reflection.[19] The UAE has not experienced a clear battlefield defeat like France in 1870, which catalyzed substantial amounts of political will for radical adoption, but the UAE has embarked on a regional influence project in response to persistent external pressure from Iran and extremist non-state actors. Since the UAE economy was modernizing rapidly in the ten or twenty years prior to military reforms intended to adopt the system of military reflection, whether UAE adoption is steady as opposed to fitful, as in Turkey, will demonstrate the importance of human capital to adoption.

My expectation is that human capital in the wider economy is critical to adopting the system of military reflection, but what kinds of evidence would substantiate this? One challenge is that the UAE is very early on in its bid to adopt the system of military reflection, and perhaps ten or twenty years need to pass before we know if it succeeds. Even if the outcome of the adoption process is as yet unknown, a fine-grained examination of NDC at this early stage can identify important decision points, or issues that could greatly hinder or facilitate the adoption process down the road.

To this end, I identify three questions that the case study of the UAE National Defense College should answer in the affirmative if human capital is critical in the adoption process. First, does participants' (or students') changed thinking at NDC influence policy making later on? If the NDC program changes the way that participants in the program identify and analyze strategic problems, but there is no discernible effect on how the government approaches policy challenges, then the human capital developed at NDC would, in effect, be meaningless for adoption of the system of military reflection.

Second, does NDC engage the country's environment for strategic dialogue? Wherever strategic decisions are made, be it in open fora or in closed councils, if NDC faculty, students, and alumni are involved in shaping the dialogue, then it would suggest an important role for human capital. Furthermore, if the environment for strategic dialogue expands or becomes more open to new ideas, it would suggest an important role for human capital in the adopting process.

Third, does NDC demonstrate continued improvement in developing the system of military reflection? If processes at NDC are geared toward cultivating each of the three elements—institutional introspection, a capacity for critical analysis, and stable civil-military relations—then it would suggest that the larger Emirati military and government organizations value the process of adoption. In particular, if the proportion of civilian and military students becomes grossly imbalanced, it could keep NDC from contributing to stable civil-military relations. If phenomena like grade inflation, intolerance for opposing views, faculty retention or recruiting problems, or a lack of curriculum development become prominent, then it would hinder the development of institutional introspection and a capacity for critical analysis. The way that College leadership address such issues, as they arise, would say a great deal about how much they, and the UAE government as a whole, value human capital in the adoption process.

The answers to these three questions have implications for adoption theory, as outlined in Table 4.1. If the answers to the three questions are "yes," then the UAE case would demonstrate the importance of human capital in adopting the system of military reflection. If the answers to these questions are "no," and the UAE also fails to continue adopting the system of military reflection, then it would likewise demonstrate the importance of human capital (by its absence).

Mixed answers to these three questions would also give insight. If the answers to the first two questions (influencing policy and engaging in strategic debate) are "no" but the answer to the last question (continued adoption) is "yes," then it would demonstrate that human capital is unnecessary for adoption. In other words, adoption theory would be disconfirmed: the system of military reflection can be bought. If the reverse is true (the first two answers are "yes" and the last is "no"), then adoption theory would be disconfirmed, but in a different way: it could mean that generating the political will to underwrite reforms in military thinking requires more than a simple desire for regional influence (UAE) or a persistently challenging external

Table 4.1 The UAE Case and Adoption Theory

Q1. Policy influence?	Y	N	N	Y
Q2. Strategic engagement?	Y	N	N	Y
Q3. Continued adoption?	Y	N	Y	N
Implication for adoption theory	Adoption theory confirmed	Adoption theory confirmed (assuming failed adoption)	Adoption theory disconfirmed (human capital unnecessary)	Adoption theory disconfirmed (greater political will necessary)

environment (Turkey). It could require a collective sense of defeat leading to national introspection, as occurred in France or Prussia.[20]

A BIRTH, WITH COMPLICATIONS

The idea for an NDC has simmered for some time in the UAE, as early as the mid-1990s. In 2011, the UAE published its *Vision 2021* document, a ten-year plan to bring the country into the modern age, with a focus on developing human capital, sustaining infrastructure growth, building a knowledge-based economy, and reducing dependence on oil.[21] The UAE's leadership saw the country playing a stabilizing role in the politics of the region, and they saw the establishment of an indigenous senior level defense college as an important part of that regional role.

By 2012, then, the time was ripe to give birth to the NDC. Not only would an Emirati defense college prepare local military officers and civilian officials to handle the burgeoning tasks of this expanded regional role, but it would also mark the UAE as a leader in senior level defense education in the Gulf. Jordan, Saudi Arabia, and Egypt have well-established defense colleges, but those are largely closed to outsiders.[22] The UAE's NDC was a departure from this model, since it was intended, eventually, to attract students from around the region, and even the world.

Leaders of the project cast around for a suitable partner. They realized that they could not do it alone, since the knowledge to run a senior level defense college did not exist in the UAE. The UAE has had a Joint Command and Staff College since 1992, and there were proposals to model the defense college on that, but project leaders wanted the best the world could offer, so they turned to the United States and the United Kingdom. After a process of deliberation among the senior UAE military leaders, the United States became the preferred partner for the UAE to build its defense college. UAE officials saw greater potential in the American model, which focuses on Socratic questioning and experiential learning in a small group format.

After the decision to adopt the American model, Emirati officials quickly identified the US National Defense University as a desired partner, given its high regard in PME. Since NDU is a conglomerate of different schools, however, a series of negotiations ensued over which school or agency at NDU would liaise with the UAE. In an environment of shrinking resources, a number of NDU schools were reluctant to devote the time and effort to support the UAE's project, but officials eventually decided that the Near East South Asia (NESA) Center for Strategic Studies would be a good fit, given its mission to develop cultural and educational relationships with governments, and especially militaries, in the region.

By late 2012, with courses projected to begin in fall 2013, officials decided that the most expedient way to support the new NDC was to use a foreign military sales contract, which the United States typically uses to sell military hardware abroad, to hire and support the necessary personnel. The new faculty would be employed by a small US defense contractor that already had a contract to provide services to NESA. The UAE would fund the contract, and the US government would essentially loan the faculty to the UAE to run NDC. This FMS arrangement created a system with at least five different approving authorities for many hiring and resourcing decisions, but the system worked quickly, at least at the outset.

In late 2012, hiring faculty was not the most difficult challenge facing NDC's Emirati leadership. The American dean that NDC hired to lead the faculty resigned, for personal reasons, so Emirati leadership scrambled to find a replacement. In February 2013, NDC hired John Ballard, a former dean at the National War College. He played a leading role in hiring new faculty, the first three of whom came on board in June 2013. These new faculty members and the dean worked furiously over the summer to write curriculum, set academic policies, and recruit more faculty to fill the complement of twelve faculty slots that the FMS case allowed.

By late August 2013, this work had borne fruit. The first three months of curriculum was written, participants began to arrive, and three more faculty signed on, with three more slated to arrive later in the year. The General Headquarters (GHQ) of the Armed Forces also assigned one of its senior advisors, who holds a PhD in political science from the University of Michigan, to teach on an adjunct basis at the College. The faculty and staff were determined to provide a quality education to incoming participants.

Despite this commitment, some challenges became evident, even before classes began. Some FMS faculties were based in Abu Dhabi, but others were based in Washington DC, so they would fly over when their turn to teach came up. This resulted in Abu Dhabi-based faculty shouldering the lion's share of the day-to-day administrative responsibilities for the faculty. Given the unusually large number of approving authorities built into the FMS contract, faculty sometimes found it hard to get clear answers on housing, salary, and other resourcing questions. Perhaps most importantly, the College was not certain if all thirty participants it expected to arrive would be released from their government positions for a year, and the curriculum for the second half of the year was not yet finalized. All the same, faculty and staff looked forward to putting their hard work into practice in the classroom, and the FMS contract produced a faculty with a useful mix of traditional academic and military backgrounds, ideal for an MA program such as NDC's. When students began to arrive, the sense of hope outweighed feelings of uncertainty.

RAISING THE BAR, THEN SETTLING IT

The College exceeded expectations in its first year. All thirty participants arrived, although a handful began the course later than the others. A number of them, mostly civilians, still performed duties they had held before arriving at the College, but they were able to dedicate the necessary time and effort to complete the coursework. These were often the strongest students. In fact, FMS faculty members routinely expressed pleasant surprise at the quality of classroom discussions.

Curriculum issues and faculty-based issues did not have a significant impact in the second half of the year. The new faculty, those hired in August and throughout the year, dove into curriculum writing and delivered thoughtful courses geared to the UAE's specific strategic environment. In terms of teaching, whether faculty were based in Abu Dhabi or Washington DC did not matter, since DC-based faculty exerted significant effort to pull their weight while on the ground in Abu Dhabi, and to engage with students over email and telephone while in DC. Overall, the curriculum was coherent and the faculty worked together with support staff to provide an education adapted to the strategic needs of the UAE.

Feedback from the first year was overwhelming and positive. Both the Prime Minister of the UAE and the Crown Prince of Abu Dhabi attended events at the College during the first year, and a number of senior officials, including ministers, spoke with the students at length about the strategic challenges facing the country. Three of the military officers who were students on the course were promoted from colonel to brigadier general, and a number of the civilian students received promotions in the months after graduation. Some members of the College community also overheard expressions of surprise, from senior Emirati and US officials, that results had been as positive as they were.

In the second academic year, 2014–15, the College sought to build on this success. Two more faculty members came on board, and the College increased its cohort size to thirty-six, split into three syndicates instead of two, as it had been during the first year. This created smaller syndicates and contributed to more engaging classroom discussions. In order to increase the cohort size, the College accepted more junior civilian applicants than for the first year, but these participants did not fall behind their first-year counterparts in terms of academic quality.

The dean and the faculty also decided that they would maintain the broad structure of the curriculum, such as course descriptions and credit allocations, they had developed in the first year, in order to focus energy on improving the curriculum, instead of changing it. As is to be expected in any faculty, there was not perfect agreement on the broad structure of the curriculum, but all agreed that freezing the structure for the first three years would be in the best short-term interest of the College's teaching mission, so fundamental curriculum questions were essentially tabled until year four of the program. This allowed faculty, in year two, to sharpen the courses they had already written. Almost no teaching assignments were changed in year two, either, in order to focus on instructional quality in the MA program.

With respect to academic achievement, the results in year two were similar to those in year one. Alumni from both of these cohorts have expressed deep gratitude for the way the MA course challenged their thinking and forced them to consider alternative ideas to their own, both in the classroom and through research projects like the year-long capstone project. One second-year alumnus, in reflecting on his capstone project and the serious disagreements he had with his faculty advisor, admitted that the experience made him better at his job as a military officer. In the months after the second year graduation, a number of civilian and military alumni received promotions, as after the first year.

That said, the second year also produced signs of trouble for the long-term strength of the College. Fewer senior leaders attended events at the College during the second year, although the Prime Minister and Crown Prince of Abu Dhabi both came. Elements of the College also began to push for the creation of a PhD program, something that other elements of the College were not uniformly behind, arguing that it would divert resources from the MA program and that the research environment was not developed enough to support a PhD program. Despite these concerns, leadership pushed ahead, and the faculty developed curriculum for a PhD program. Perhaps most importantly, for a variety of reasons, four out of the eleven FMS faculty members left after the second year, which signaled a potential challenge with faculty retention and a loss of institutional memory due to faculty turnover.

In the third year, the College continued to experience both successes and challenges. Course participants continued to impress with the perceptiveness of their insights, and the MA curriculum remained coherent, relevant, and sufficiently challenging for the most advanced participants. Two new faculty joined NDC, which is less than the four that left after the second year, but it demonstrates that the College was aware of the faculty manning issue. Two other faculty members left the College in the third year, although leadership also identified a handful of Emirati faculty members with the ability to teach courses (though not write them, yet).

Despite these faculty retention challenges, the College improved a number of its support functions in its first three years of existence, especially in classroom technology and student services. Other support functions, such as library services, outside speaker coordination, and faculty research support, lagged well behind.

In its fourth year, the College experienced increased tension over support for faculty research. New faculty hires stabilized teaching loads, and policies were proposed to regulate research; but these efforts did not address a key stipulation of accreditation, which was to bolster faculty research, which is an essential element of creating an intellectual environment hospitable to introspection and critical analysis. In this regard, foreign and FMS faculty members generally realize that their roles at NDC are temporary, but support for faculty research has been uncertain, and granted only on a case-by-case basis (although funding is generous when granted). This has forced faculty to accept reduced research productivity while waiting to see if the research support environment improves. This dynamic has implications for one of the two big "ifs"—accreditation by a US body—and also for faculty retention, since faculty will be less likely to remain in an unstable research environment. Faculty turnover risks a loss of institutional memory and the understanding of what it takes to meet the UAE-specific needs of participants at NDC.

Based on increased faculty numbers and a push to demonstrate the capacity to expand cohort size, the College increased the number of syndicates from three to four in year five. As of this writing, this increase has improved classroom discussions, since smaller syndicates give more participants the opportunity to speak (cohort size has not yet increased). In addition, College support services, in particular facilities management and information technology, are top-notch, and would be the envy of most schools in the West. Most importantly, however, the College has demonstrated a continued capacity to place alumni well—the other big "if"—especially the military alumni. In fact, two alumni were placed on a committee to evaluate research projects in the armed forces, and using the research skills acquired at NDC they conveyed to senior military leadership how the armed forces could improve research activities in a systemic way. In short, NDC is answering the alumni placement "if" in a most positive way.

Progress on the other "if"—US accreditation—seems more tenuous. The tension over the role of research and its accreditation implications threatens the long-term health of the College. This tension first manifested itself in a conflict over the PhD program in the middle of the third year. At that time, the College welcomed an outside accreditation team for its PhD program, and one result was a push to give more support to faculty research. As of this writing, in the fifth year, the faculty's ability to use outside funding to conduct research is somewhat constrained, although this does not differ substantially from practice in US professional military education institutions. Funding for conference attendance also still operates on a case-by-case basis, although there are indications that this may be changing. On the accreditation "if," then, progress is not nearly as decisive as for the alumni placement "if."

When it comes to faculty recruitment, the College has a robust capacity for identifying the best among a group of candidates, but little reach to expand the applicant pool because faculty are less engaged than they could be with research. In addition, a lengthy hiring process has sometimes led to faculty not actually coming to the College, even though an offer has been extended. In an environment where faculty retention and recruitment are such a challenge, it is difficult to strike a balance between research support and teaching quality.

If the College does not have enough faculty members to teach its courses, research will inevitably suffer, since the primary mission of the College is to teach the MA program. The College partly alleviated this challenge by postponing the start of the PhD program for at least two years, with a projected start in 2018, but the increase to four syndicates in year five still meant an overall increase in the teaching workload. If any more faculty choose to leave, for whatever reason, this workload problem will deepen, which would itself create a disincentive for faculty to remain, especially if they are interested in building a research program recognized by their peers in the field. Ironically, the College needs these same research-focused faculty to build a strong, credible PhD program in the future and to provide a challenging yet relevant MA program. In the absence of a systemic fix to this faculty retention and recruitment challenge, the College's commitment to supporting faculty research lacks credibility and, more importantly, launching new academic programs could harm the quality of the MA program.

A separate, Arabic-language War Course has also been proposed, for senior officers without the English skills to succeed in NDC's MA program. This could also present challenges to the College mission to instill patterns of strategic thinking in its participants and help the UAE adopt the system of military reflection. War course participants would consist almost entirely of military officers, and the course duration would be a few months instead of a year, so there is a chance that it could siphon officers from the original, English-language National Defense Course. To the extent this occurs, it would impact the College's ability to cultivate stable civil-military relations and instill institutional introspection across the UAE defense establishment, especially if the war course does not adopt the Socratic, open style of discussion now present in the original National Defense Course. At the same time, the War Course could exceed expectations, just as the National Defense Course did in 2013 and 2014, so only time will tell if challenges or successes win out.

The College is thus at an inflection point between challenges and successes. While participants are not as well-read as many of their Western counterparts in military education, they boast much more experience in developing strategy and making policy. They see the immediate impact of learning the material, because so many alumni are placed in meaningful policy making positions soon after graduation. The trend of alumni promotions continued after the third year, although the College has yet to conduct a systematic survey of alumni to ascertain how widespread this trend is. As the College continues to attract highly-experienced participants for the MA program, the success in placing graduates will continue to magnify NDC's positive reputation across the UAE government. Every indication is that NDC will continue to attract solid MA participants: applicant interviews have been conducted earlier every year, and the College provided MA participants for the fourth and fifth cohorts with summer reading before they arrived in August. These are important markers of success.

The ability to continue producing quality graduates for the UAE government may hinge, however, on faculty retention and recruitment. The College relies on Western faculty, essentially imported human capital, to maintain the quality of its MA program, and the same will be the case for the PhD program when it begins, though likely not for the War Course. If the College commitment to faculty research lacks credibility because of workload issues, then it would threaten the balance between faculty with academic and military backgrounds, which has been a strength of the College thus far. The College in particular, and military education in general, needs traditional academics to leaven the instruction with alternative views. The decision to include more traditional academics on the faculties of US military schools in the 1990s raised the quality of instruction,[23] so maintaining this balance at NDC would help build the human capital of future Emirati strategists and, ultimately, contribute to adopting the system of military reflection.

To the College's credit, the leadership of NDC seems to recognize that the College faces a challenge with faculty retention and recruitment, even if a systemic fix has not yet been presented. One positive sign of NDC's commitment to faculty research is that it allowed this chapter to be written, even though it is critical of the College in some ways. This openness to alternative viewpoints is the essence of human capital creation, and will contribute to NDC's positive academic reputation as it educates future Emirati leaders.

HUMAN CAPITAL MATTERS: THREE QUESTIONS

This chapter poses three questions that, if answered in the affirmative, would suggest the importance of human capital in the process of adopting the system of military reflection. The first question is whether participants' changed thinking at NDC influences policy making later on, a question that requires evidence on whether participants' thinking changes at NDC at all, then on whether alumni subsequently influence policy.

Evidence suggests strongly that NDC alumni influence policy making in the UAE. Of the thirty alumni from the first cohort, six received promotions in the two years since graduation. Promotion demonstrates that the government values the contributions these alumni make, but it does more than that: it puts alumni in influential positions. For example, civilian graduates from NDC have been asked to establish a new Emirates Diplomatic Academy for educating diplomats from the Ministry of Foreign Affairs; to direct the National Media Council; to advise the National Security Council; and to direct the Americas Directorate at the Ministry of Foreign Affairs. Military graduates have been asked to command the UAE's task force in Yemen, command the Medical Services Corps, and administer the UAE's military industrialization program. Alumni also influence subordinates and colleagues at their home agencies to attend NDC, which suggests that alumni are well-regarded enough that their agencies are willing to relinquish their most valuable employees for a year of schooling at the College.

Evidence also suggests that participants' thinking changes while they are at NDC. Faculty regularly comment on how participants' analytical capabilities improve during their year at the College, especially on the capstone paper, a year-long policy analysis project that is required to receive the Master's Degree.

Evidence of changed thinking also comes from participant course grades. Unlike most civilian university schedules, which follow a concurrent course model, NDC (like many military schools around the world), follows a block schedule, where one course is taught at a time, instead of at the same time as other courses. Participants at NDC thus receive a course grade every two to four weeks throughout the year. I examined the grades of a sample of thirty participants, chosen to be representative of the typical mix of civilians and military officers. Figure 4.1 shows the results.

One common misconception is that military officers systematically underperform compared to civilian participants, but these data indicate that this is not the case. In some courses, military participants outperform their civilian colleagues. In addition, course grades exhibit a systematic increase over time. As the year progresses, the participants' ability to engage in critical analysis improves, as suggested by the across-the-board increase in grades over time. This evidence, along with the clear trend to place graduates in important policy making positions, suggests that the answer to the first question is "yes." Participants' changed thinking influences policy.

The evidence is slightly less clear for the second question: does NDC engage the country's environment for strategic dialogue? The main challenge to observing evidence to answer this question is that most meaningful policy discussions in the UAE take place out of the public eye, in the ruler's *majlis*, or council. That said, current and former participants often speak about discussing issues of strategic importance with senior national leaders, and

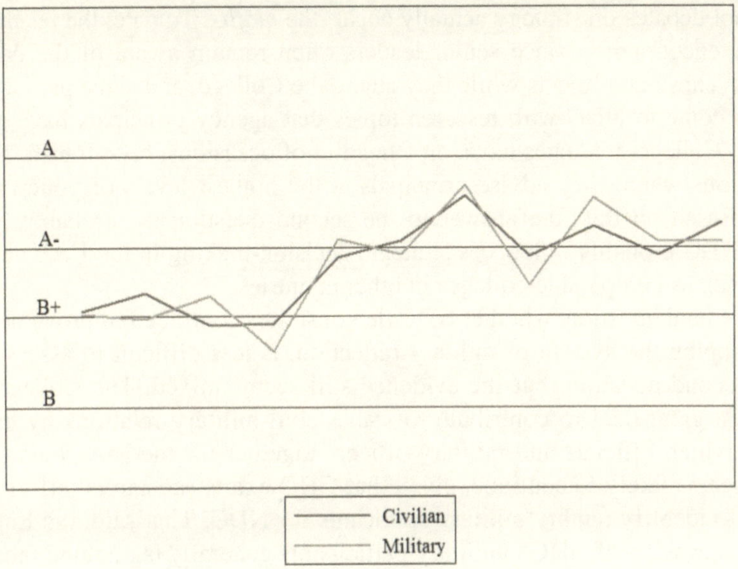

Figure 4.1 A Representative Sample of Student Performance at NDC. *Data courtesy UAE National Defense College.*

NDC faculty and leadership often contribute to periodicals that focus on strategic questions, such as the GHQ-produced *Nation's Shield* and the College-produced *National Defense*. Every year, senior leaders receive copies of the *National Defense* magazine, in which participants' writing features prominently, and every year senior leaders have requested a list of capstone paper titles, along with their abstracts. Two faculty members were even asked to brief the National Security Council. Without question, the College is making the attempt to influence the country's strategic dialogue, and there is some evidence that this attempt is having an impact.

But it is hard to say how much impact these efforts have. Public lectures on issues of strategic importance, such as those at the Emirates Center for Security and Strategic Research or New York University Abu Dhabi rarely challenge official government thinking, and with one or two exceptions Western faculty's writing is rarely printed in Arabic. In fact, one faculty member prepared an op-ed in both English and Arabic, for publication in local newspapers. The English version was readily published, but the Arabic version was seen as too critical to be published, suggesting that non-Emirati faculties are, at the very least, discouraged from engaging in the Arabic-language strategic discourse.

Given the embryonic state of public strategic debate in the UAE, the real question is whether NDC can influence the dialogue in the forum where vibrant debates on strategy actually occur: the *majlis*. To date, the results are fairly encouraging, since senior leaders often remain aware of the participants' capstone projects while they attend the College, and since participants often come to NDC with research topics that agency principals have given them. College graduates have, on a handful of occasions, been transferred to positions where they advise principals at the highest levels of policy making. Based on this, the answer to the second question is "probably, yes." The NDC probably influences strategic decision-making in the UAE at least as much as comparable colleges in other countries.

The final question, whether NDC demonstrates continued improvement in developing the system of military reflection, is less difficult to assess than the second question, but the evidence still seems mixed. The College has done a great deal to contribute to stable civil-military relations by bringing civilian officials and military officers together for the MA course, and this seems likely to continue, given that GHQ exerts substantial effort each year to identify quality military applicants for NDC. That said, the English language skills of NDC's military participants generally lag behind those of civilian participants, suggesting that it could be harder as the years go by for GHQ to find suitable candidates for NDC's English-language course. Any significant reduction in the proportion of military participants could harm NDC's ability to foster stable civil-military relations, although, at the same time, there has been almost no disagreement among the participants in what the proper roles of civilian and military agencies are in national defense, and Emirati military officers demonstrate a strong professional ethos of loyalty to the country's rulers. All told, NDC has contributed to stable civil-military relations in the UAE, and it seems likely to continue to do so, albeit with some challenges.

The College has also worked hard to cultivate introspection, at least within the walls of the institution. Faculty, participants, and staff take the College's non-attribution policy seriously, and discussions in the classroom are generally open and wide-ranging. On a number of occasions, former participants have remarked how much they enjoyed the freedom to think that the College offered. Almost all alumni comment on how much they would like to come back, which is a testament to the intellectual haven that NDC leadership have created. Faculty have also enjoyed wide latitude in creating courses and establishing the basic educational policies of the College, such as grading, appeals, and curriculum review, which is different from the institutional constraints that faculty at US military schools sometimes face. Furthermore, a new coordinating council of College principals works on establishing and implementing policies on a systematic basis, which is at odds with the decision-making based on status and personal connection that Arab militaries have often been accused of using.[24]

This introspection does not extend into larger society, however. In order for this introspection to become institutionalized, Emirati society would have to evolve practices that are not welcome, such as due process based on individual rights, freedom of speech, and active political participation. These changes are extremely unlikely to happen. Rather, Emirati society uses the *majlis* as a mechanism for introspection, with tribal leaders and the well-connected conveying negative reactions to government policy to the country's leaders. If the College itself is to generate institutional introspection outside its walls, then it will have to do so by influencing the mindset of its alumni before they diffuse throughout the *majlis* system. This is a long process. The College, then, has certainly cultivated introspection to the extent that US military schools have, even if it will likely be confined to the walls of the College for the near future.

The capacity for critical thinking at the College is not as robust as institutional introspection. It is one thing to reflect and suggest new ideas, but it is quite another to critique the worth of others' ideas. As participants work their way through the MA program, their critical thinking skills improve, but participants, whose native language is Arabic, not English, often find it difficult to digest the number of pages per night that would allow them to ask incisive questions. In the question and answer sessions following lectures, which are given every day, faculty are asked to defer to participants, which robs participants of the opportunity to see how critical, constructive questions are formed. Certain ideas are often left unchallenged, such as animosity toward Iran, the status of the Muslim Brotherhood, or attitudes toward *shi'a* Muslims. Every year, a handful of participants choose capstone topics considered politically safe, and thus well-worn, which has in turn forced faculty to take measures to ensure that topics are not repeated year after year. The College on occasion exercises a security review prerogative on faculty publications, which can chill how much faculty push the envelope of their critiques.

When it comes to building a capacity for critical thinking, the College is fighting an uphill battle, but there are some encouraging signs of progress. On specific policy issues, some participants weigh options against evidence in their capstone papers, instead of using evidence to reinforce forgone conclusions. Emiratis are also generally open to critiques that are fair and humble in tone, as long as they do not cross certain red lines, like demeaning national leaders or insulting Islam. They are accustomed to listening to ideas different from their own, given that over 200 nationalities are represented in the country. The UAE has also proven itself more willing than other Gulf states to allow foreign religions to practice in the open, as long as the ban against proselytizing Muslims is observed. The College in particular, and Emirati society as a whole, seems to tolerate opposing ideas to a greater extent than many of their Arab counterparts. In terms of a capacity for critical thinking, they may be similar today to where Prussia was when it began to adopt the system of military reflection in the nineteenth century.

Given the mix of evidence, it is difficult to answer the third question—on continued development of the system of military reflection at NDC—with absolute clarity. Civil-military relations appear to be very stable, and the College is a haven of reflection, but so far it boasts only a limited capacity for critical thinking. Thus, the answer to the third question is "possibly, yes," while the answers to the other two are "yes" and "probably, yes."

The answers to the three process-tracing questions are qualified, but they lean affirmative, suggesting that the UAE recognizes the importance of human capital as it adopts the system of military reflection. If the UAE thought it could simply buy the institutions of a modern military, as it seems to do with many capital infrastructure projects, then it would not be as concerned with developing the human capital of future Emirati decision makers. It would not have endowed its NDC with high-level official sanction, nor would it have removed some of its most highly-skilled officials from government service for a year to attend the College. It would not have devoted the resources it has to building NDC's support functions and bringing capable faculty from halfway around the world to teach there, nor would it have empowered alumni to influence decision-making at the highest levels of government. The College may have more than its fair share of challenges, but it represents an important commitment to developing human capital as the UAE reforms its military.

An important caveat is in order. The NDC does not represent the sum total of the UAE's efforts to adopt the system of military reflection. Such efforts occur at many levels, especially in operating forces like the Presidential Guard. The push to adopt the reflection system also began before NDC was founded, possibly even a decade earlier. At the same time, NDC is an important part of this effort, and it constitutes a microcosm for adoption in the country as a whole. The same status quo forces that present challenges to NDC's development also confront other organizations seeking to adopt aspects of the system of military reflection. The process of adoption in the UAE is thus at an inflection point, like NDC, and it turns on one critical factor: human capital.

HUMAN CAPITAL: THE PIVOTAL LINK

The UAE case demonstrates the importance of human capital in adopting the system of military reflection. It is extremely wealthy, but with poor human capital, so its efforts to develop the National Defense College suggest that human capital is at least as important as material capital in the process of adopting the system of military reflection. The UAE recognizes that the reflection system cannot simply be bought.

The UAE case also conforms to the expectations of adoption theory. It has not experienced a catastrophic battlefield defeat, although its bid for regional influence motivates its desire to adopt the system of military reflection. At the same time, its economy is modern in many ways, albeit modernizing in others. Given this mix of factors, adoption theory predicts that the UAE should reflect a steady adoption process. It is still too early to judge the ultimate success of the UAE's adoption endeavor, and thus determine whether its path to adoption is steady or fitful, but the initial evidence is suggestive: NDC, for example, is not going away anytime soon, but it is also not going to become world-class overnight, and it has serious work to do if it wants to attain such heights. By comparison, the US Army War College was in operation for some fifteen or twenty years before the nation truly began to see it bear fruit.[25]

The UAE case also presents puzzles for the competing explanations of adoption. It is not democratizing, at least not by any metric used in Western political science scholarship. It is not moving closer to the West culturally, but rather seeking to forge its own way. Its principal external threat amounts to Iran occupying three mostly uninhabited islands in the Gulf and intervening in the domestic politics of other Arab states in the region.[26] But this state of affairs has existed for over forty years, while the UAE launched its bid to adopt the system of military reflection in the last five to ten years. If democratization, cultural distance, and international security explain adoption of the system of military reflection in other cases, they do not do so in the case of the UAE.

The main strength of the UAE case, however, is not in making inferences about which factors lead to adoption, because the UAE is still early on in its adoption process. The real power of the UAE case is in demonstrating how the development of human capital plays a role in that process. If ever there was a case where the presence of human capital should "not" matter for adoption, the UAE is it, but instead this case suggests that human capital is a pivotal link between a modern economy and adoption of the system of military reflection. A country can have all the money in the world, but if it does not have human capital it will not get much closer to the system of military reflection.

The evidence presented in this chapter still leaves one question unanswered: can human capital influence adoption of the system of military reflection if it is confined to the schoolhouse? A real possibility exists of NDC becoming a place where participants and faculty think splendid thoughts, but then those thoughts never leave the building. Some signs indicate that this is not happening—the promotion of alumni and recruitment of highly-placed participants—but other signs, more worrying ones, indicate that it is—little engagement by College faculty in the public strategic debate and the paucity of critical policy analysis in society writ large. In other words, the rate at which knowledge of strategic affairs is accumulating in the wider UAE society is much slower than the human capital accumulation rate at the College. If internal College debates never translate into public discussions, and if strategic decision-making continues to happen mostly behind closed doors, can the Emirati polity develop the human capital it needs to sustain the system of military reflection?

Other cases of non democratic adoption indicate that the answer is probably "no." In Prussia, which experienced a disastrous battlefield defeat before adopting the system of military reflection, the strategic direction of the country was debated in the press, especially after the 1806 defeat. Prussia produced some of the most highly-regarded military officers of the nineteenth century, including Carl von Clausewitz and Helmuth von Moltke, as well as Otto von Bismarck, whose claim to fame as a strategist came as a civilian statesman, not a military officer.

Russia, which did not experience a battlefield defeat before it began to adopt the system of military reflection in the early twentieth century, has experienced vocal opposition voices throughout its history. Some of these voices have been silenced under mysterious circumstances, especially of late, and the Russian state has historically been heavy-handed in dealing with opposition, but a vibrant culture of public strategic debate still exists in the country. The Russian military academy system developed ground-breaking concepts of operational art and deep battle in the interwar years, which may have contributed to the German concept of *blitzkrieg*, and certainly influenced American thoughts on combined arms maneuver later in the twentieth century.

Emirati officials clearly understand the need to develop human capital in the country's future strategists, but there seems to be little stomach, as yet, for a robust and public discourse on strategic affairs that takes a detailed look at all options facing the country. Emirati policy making typically follows a more private path. While two cases, Prussia and Russia, do not decide definitively whether a critical, public strategic discourse is necessary to sustain the system of military reflection, they do illustrate the challenge that the UAE faces in forging its own path. The UAE has the wealth to succeed in its bid for adoption, but only time will tell whether societal human capital and an open strategic discourse are necessary for adoption as well.

NOTES

1. UAE 2012; Bell 2013; NDC 2016.
2. Salem 2012; ECSSR n.d.; Abdulla 2014.
3. Lynch 2011; 2012.

4. Ibish 2017.

5. Barnett 1998.

6. McGregor 2014. The UAE Armed Forces have launched military operations abroad in Lebanon (1976, 2001, 2006, 2009), Kuwait (1991, 2003), Somalia (1993), Bosnia (1996), Afghanistan (2003, 2006, 2007), Pakistan (2006), Yemen (2008, 2015), Libya (2011), and Yemen (2011, 2014).

7. Chandrasekaran 2014.

8. McNeill 2016.

9. Abu Dhabi University 2016; JCSC 2016.

10. Toronto 2014.

11. NDC 2016.

12. Heard-Bey 2005. Other countries, like the United States, exhibit similar family clustering in military service. See Janowitz 1971; Cohen 1985.

13. NDC 2013.

14. Ibish 2017.

15. AAA 2012.

16. An extensive surveillance state could have implications for how a country adopts the system of military reflection, as intrusion into individuals' private lives can empower conservative forces seeking to maintain the status quo. It is unclear whether this is or could happen in the UAE. On the one hand, the UAE's leaders are generally forward thinking and understand the necessity to change for the future. On the other hand, the UAE—and Abu Dhabi, in particular—is a strongly traditional, conservative society.

17. FCSA 2016.

18. Naidu, Nyarko, and Wang 2015.

19. Bennett and Checkel 2015.

20. Alternative mixed answer permutations are both less likely and less interesting than the "YYN" and "NNY" scenarios. The answers to the first two questions, on policy influence and strategic engagement, are probably correlated with one another, since there is almost no open public policy debate to speak of in the UAE. Given the emphasis on centralized policy making, generally behind closed doors and among a small group of people, strategic engagement by the College is likely to have less of an impact on actual decisions than the policy influence of alumni and their changed thinking. Thus, the remaining two possibilities, "YNN" and "NYY," would likely conform to the expectations of the "YYN" and "NNY" scenarios, respectively. I include all three process-tracing questions, however, in case future researchers choose to apply this methodology to cases where strategic engagement has a more decisive role in policy making.

21. UAE 2016.

22. Oman also established a defense college in 2013.

23. Watson 2007.

24. Pollack 2002; Talmadge 2015.

25. Schifferle 2010; Matheny 2011.

26. Mattair 2005.*Chapter 4*

Chapter 5

Conclusion

Making Military Success Last

The most successful states in the twenty-first century will have the intellectual wherewithal to turn military prowess into strategic success. States might put trust in the good faith of allies or in the sophistication of modern weaponry, but these will not guarantee success on the battlefield. To turn the raw materials of national power into strategic success, states must rely on knowledge.

Some states have acquired this knowledge, which I call the system of military reflection, but others have not. Why? Many states, especially from the West, spend millions of dollars and thousands of hours every year trying to teach, reform, and train foreign military forces around the world. Thousands of military officers learn the art of war, yet the forces they lead sometimes fail on the battlefield.

War is the province of chance, as much as reason and will,[1] so nothing can guarantee victory, but applying the modern system of tactics certainly improves the odds.[2] This system of tactics does not become instituted in a military force because foreign trainers and educators come to teach course after course, year after year. It relies on a deeper system of thinking, one marked by an aspiration to excellence. Military forces like those of the United States and its NATO allies, Russia, Australia, New Zealand, Israel, a number of Latin American militaries, and perhaps China and a few other countries have acquired this knowledge, which is defined by stable civil-military relations, institutional introspection, and a capacity for critical thinking. A number of other states, especially those in the former Soviet sphere and the Arab world, are in the process of adopting this system of reflection, because they know that their security in the future depends on it.

What are their chances of acquiring these patterns of thought? It depends on their ability to modernize the economy, since a modern economy generates the human capital and wealth necessary to sustain the system of military reflection over the long term.

But battlefield experience also matters. States that begin reforming the military on the heels of a catastrophic defeat, like Egypt in 1967, are likely to have the political will to make dramatic reforms at the outset, while those that merely experience external pressure, such as Turkey, or the UAE in its bid for regional influence, are more likely to take a more measured approach to reform. Combined with whether the economy is modern or modernizing, this results in four pathways to adopting the system of military reflection: radical (France), steady (UAE), fitful (Turkey), or front-loaded (Egypt, Prussia).

I have presented evidence that spans from the general to the specific. A global database of military schools, stretching over 200 years, suggests a two-stage pattern of adopting the system of military reflection. States with greater security concerns are more likely to reform the military, but wealthier states, with higher rates of human capital, are more likely to have staff colleges, war colleges, more military periodicals, and higher military expenditures per soldier than poorer, less educated states. The statistical evidence also casts doubt on three competing explanations for adoption. Being a democracy (Regime Type Hypothesis), having international security concerns (in isolation; International Security Hypothesis), and cultural proximity to potential models of military excellence (Cultural Distance Hypothesis) are not strongly correlated with indicators of the system of military reflection.

The case studies illustrate the four pathways to adoption, and in doing so strengthen the main contention of adoption theory, which is that the timing of battlefield catalysts and economic modernization determines how militaries learn. Militaries learn on the battlefield, but they also learn by absorbing human capital from society. Egypt, Turkey, and France launched serious reform efforts in response to external security pressure, but these reforms did not result in adoption of the system of military reflection until after their economies had substantially modernized, which weakens the argument that international security concerns drive adoption. Economic modernization helped institutionalize adoption in Prussia, and appears to be doing the same in Egypt. Egypt, Turkey, and the UAE are adopting the system of military reflection, even though they have kept themselves at arm's length culturally from the West, which weakens the cultural distance argument. Prussia adopted, and Egypt and the UAE are adopting, the reflection system in the absence of democratization, which further undermines that argument. All five case studies demonstrate that states take a more nuanced approach to adoption than merely mimicking the best militaries in the system; adopters follow a steady, fitful, radical, or front-loaded approach depending on the timing of catastrophic battlefield defeats and economic modernization.

The UAE case also illustrates the importance of human capital in adopting the system of military reflection. Wealth and human capital are the links between a modern economy and adoption, but the UAE experience suggests that the system of military reflection cannot simply be bought. The UAE economy is modern in terms of wealth, but poor in human capital. It is also a non-democracy, culturally distant from the West, with relatively mild external threats, although it has begun a military reform program seemingly aimed at regional influence. It also exhibits a concerted effort to improve human capital as part of this process, which suggests that human capital is at least as important as wealth in adopting the system of military reflection. In fact, the UAE as a whole may be in a race to build its human capital base before flagging oil prices leave it without the resources to truly modernize its economy. UAE leaders, it seems, recognize the importance of human capital.

RECOMMENDATIONS FOR POLICYMAKERS AND RESEARCHERS

Policymakers and researchers alike should recognize the importance of the economy in building military power. American and other Western officials working to reform security forces around the world should link security assistance to economic development. Wealth and human capital in society can sustain military reforms and schools once Western advisors leave, and help recipient countries to break a cycle of dependence, where they rely on Western advisors because they do not have the expertise to train and educate their own forces, but then are unable to transition their own people into former advisors' roles because they worry they would not provide quality instruction.

Western governments should also be more selective in which countries they offer security assistance to. Many security assistance missions build military-to-military relationships between countries, but going beyond this basic level of resourcing should involve a calculus about which countries will most benefit from security assistance, in particular those countries that have shown willingness to reform their economies and build human capital in their societies. Western governments could also offer challenge grants to potential security assistance recipients, so that potential recipients compete for the most lavish security assistance funding, with grant applications consisting of the steps the applicant country is taking—in society as a whole, not just in the military—to develop wealth and human capital. If Western security assistance is run on a dole system, with assistance a means of mere political leverage, it will contribute to the mistaken impression that recipients do not need to work that hard to establish the intellectual foundations of battlefield success.

Countries providing military aid could link security assistance to economic development by helping recipient security forces to identify and recruit the most capable individuals possible from society. Leaders of security assistance missions could also help recipient countries to develop a strategy for devoting the necessary resources to long-term reform, one that assumes a phasing out of Western funding support over time. Where feasible, Western military advisors could assist foreign security forces to focus recruitment in areas where governmental and nongovernmental economic development programs are either ongoing or have achieved development goals—as opposed to areas where little development has occurred—and could advise military leaders on the retention of the most capable military officers and enlisted personnel.

These approaches might not work in all countries. Circumstances sometimes dictate which countries will receive security assistance, as in Afghanistan. The Afghan economy is still a long way from being modern, but this does not mean that there is no hope for military reform there. Intensive external support had "some" positive effects on military reform in Ottoman Turkey, even absent a modern economy. But both the provider and recipient of security assistance should be realistic about potential outcomes. Withdrawing training and equipment support from states such as Afghanistan, before their economies are significantly developed, can undermine battlefield effectiveness. Security assistance providers should recognize that they may not be able to accomplish all their goals in a country, and recipients should recognize that security assistance is not a panacea for a deeper lack of human capital or wealth in their societies.

The development of military force is part of a larger system of societal development, so economic modernization may require reform of political or economic institutions to succeed. To build a human capital base, it is not enough simply to send students to school and expand literacy in society. Developing human capital requires a knowledge-based economy that provides meaningful incentives to innovate and produce knowledge, instead of simply consuming information. One of the singular successes of Western economies, even before many of these countries democratized, was that they began building institutions that rewarded innovation with profit. Many economies in the Middle East, and especially in the Arab world, are controlled by oligarchs that impose steep barriers to entry for small- and medium-sized enterprises. Rentier effects are also extremely damaging to human capital development, which Middle Eastern leaders must account for as they seek to expand human capital in their countries. Making the economic changes that would underpin adoption of the system of military reflection, for Arab and other Middle Eastern militaries, will create winners and losers that, inevitably, will put strain on existing political bargains.

On a more positive note, the findings of this study allow policymakers to be opportunistic. If a security assistance recipient, or potential recipient, suffers a significant defeat on the battlefield, as Iraq did in 2014, then Western policymakers could use it as an opportunity to cultivate the political will for meaningful change in a military force. Iraq has a substantially more modern economy than Afghanistan, so from that perspective the prospects of establishing the intellectual foundations of battlefield success are better in Iraq, but Western policymakers may have missed an opportunity to push Iraqi leadership to stabilize civil-military relations in the country and depoliticize the security forces, which were the main reasons that Mosul and Ramadi fell to ISIS fighters the first place.

Western policymakers intent on building partner capacity and security force assistance should move substantially beyond training and equipping foreign forces. Developing the intellectual architecture of battlefield success is a systemic endeavor, not one of getting the right training and equipment to individual soldiers. Intensive external training support was present before both failure and success in Turkish and Egyptian military reforms, but these countries did not start adopting the system of military reflection until the right economic conditions were in place. Training and equipping foreign forces may be a necessary step along the path, but the path should not end there.

Instead, more intensive focus on sharing students and information across military schools is necessary. Military schools are notoriously tight-lipped about curricula, assessment methods, and best practices in the classroom; but sharing this information in a direct, unclassified, and open way would likely improve military education around the world and deepen knowledge communities committed to peace and security in the twenty-first century. The College of Commandants is the most prominent venue for this sort of information sharing, but neither the depth of information shared nor the number of people involved is widespread enough to institutionalize best practices across national borders. Having interacted with officers and educators from military schools around the world, I can say that there are many more transnational similarities than distinctions in what is taught at military schools and how it is taught. In the information age, it is high time for military educators to overcome the knee-jerk reluctance to share information based simply on a fear of passing on national secrets. Military education around the world would benefit if there were a common knowledge center for curriculum insights, speaker profiles, best practices, and faculty job offerings.

This study also provides insights for security studies scholars, particularly with respect to military power. Understanding military power goes beyond battlefield performance. We can make inferences about the competence of a state's military by examining military education and operational planning systems, and then comparing that to battlefield performance to see if a given military succeeded based on having luck or facing a weak opponent or having better equipment or, as this study should lead us to suspect, fielding more competent soldiers. This study, in sum, offers another way to understand military power.

This study also has something to say about democratization. This process may have contributed to the development of capable, depoliticized military forces in Latin America, but this has not happened everywhere. Future research could investigate under what conditions political change could lead to more effective military forces, an important question for places like Ukraine and Pakistan, which are democratizing while fighting rebellions within their borders. This study suggests that the key to building effective military forces is human capital, which has often been a by-product of democratization, but not always. Can autocracies be just as effective militarily as democracies, given the right conditions? Future research could investigate these links more.

Future research could also take a more nuanced approach to understanding the role of security concerns in changing civil-military relations. Under what conditions, for example, will poor battlefield performance be characterized as a catastrophic defeat that political entrepreneurs can use to change how civilian and military leaders interact? Does performance against external opponents have a different effect on civil-military relations than performance against internal ones? In Iraq, for example, what would it take to construct the stable civil-military relations necessary to bring security to the country?

Finally, this study asks scholars to take a closer look at military education as a factor in security. As the quality of military education improves, does it improve security? Are states with more advanced military education systems more powerful (or do they develop these military education systems "because" they are more powerful)? This study suggests that understanding more about military education will give greater insights into security around the world, and it offers a new dataset to help scholars gain these insights. It may be that the sword is mightier than the pen, but a wise pen goes a long way to sharpen the sword.

NOTES

1. Clausewitz 1976[1831], 89.
2. Biddle 2006; Grauer and Horowitz 2012.

Appendix

Statistical Background for Adoption Theory

This appendix provides technical details for the statistical correlations presented in chapter 1, which began by establishing a link between military education and battlefield effectiveness, then explored the reasons why states might reform military education in order to adopt the system of military reflection. The academic literature described in that chapter suggested six hypotheses, and the ensuing statistical analysis led to theoretical expectations for the case studies, commensurate with adoption theory's two-stage model for adopting the system of military reflection.

MILITARY EDUCATION AND BATTLEFIELD EFFECTIVENESS

Does military education improve battlefield effectiveness? This question is unexplored in the academic literature, yet it forms part of a persistent debate on the relative merits of officer education versus field service in preparing a military force for war. In an effort to begin settling this question, I have collected data on all military academies, staff colleges, and war colleges from 1800 to the present, which allows me to answer the question, "Does having a military school impact battlefield performance?"

In previous research,[1] I described the collection of data on military academies, or schools meant to prepare officer cadets for future service as small-unit leaders in a military organization. However, the vast majority of states have schools that meet this threshold for being a military academy, so the presence of a military academy is not an effective discriminator between militaries that are effective and those that are not. What is more, a substantial proportion of an officer's intellectual development occurs after graduating from a military academy, in staff and war colleges (among other, more role-related training courses). In addition, while most militaries draw a majority of the officer corps from national academies, the United States draws the majority of its officers from Reserve Officer Training Corps (ROTC) programs on college campuses nationwide, and the US military is considered one of the most effective in the world. This, combined with the paucity of countries without a military academy, motivated me to collect data on higher level military schools.

The presence of staff and war colleges are more compelling discriminators of countries' military education systems. In 1800 only a few countries had a staff college, which are military schools focused on preparing officers to transition from tactical command to organizing the elements of military force to accomplish strategic objectives. Today the vast majority of countries have staff colleges, but few of these countries established staff colleges in the first few years of independence, so it takes an appreciable level of effort to establish and maintain a staff college. In the data set of country-years from 1816 to 2011, 34.9 percent of country-years are coded 1 for the *Staff college* variable, meaning that a country had at least one staff college in that year (and 0 otherwise).

The distinction with war colleges is even more stark. A war college is a school meant to prepare a country's current and future leaders, be they in the military or related agencies, to organize the elements of national power to achieve strategic goals and preserve national interests. The world's first war college, under this definition, was established in 1874, and only sixty-three countries today boast a war college.[2] From 1816 to 2011, 14.9 percent of country-years are coded 1 for the *War college* variable (0 otherwise).

I use these two dummy variables to assess the link between military education and battlefield effectiveness, but it is worth pointing out why I do not report results based on count variables, or the number of staff or war colleges a country has, instead of whether they have at least one of each of these. The organization of military education varies widely from country to country in many respects, but they vary less in what it takes to build that first school. In other words, the meaningful discriminator in military education is between having no school (whether staff college or war college) and having one. Having additional schools does not appear to be meaningful, either anecdotally or statistically. Statistically, using the count variable as opposed to the dummy created a larger number of outliers, skewing the results toward countries with large militaries, like the United States and China, or those that happened to have a large number of schools, like Germany and Russia around the turn of the twentieth century. Anecdotally, countries seem to dedicate the greatest amount of military reform effort to establishing the first school at a given level, and the founding of the first school is often imbued with a deep sense of national pride, as it marks the point at which a national military education system "began." After that, building the next school appears substantially easier and has less national meaning. For these reasons, I use dichotomous indicators of military schools instead of count variables.

Many academic studies of military effectiveness have used a dataset of 125 battles between 1913 and 2003 to assess the battlefield effectiveness of military forces, since the dataset codes a wealth of information, including numbers of troops and tanks on each side, air sorties conducted, and casualties suffered. The US Army originally collected this data, but Stephen Biddle and Steven Long updated and improved the data for an oft-cited 2004 study examining the links between democracy and military effectiveness.[3] Even though this data is widely cited in the military effectiveness literature, they suffer from an overrepresentation of the United States, Germany, and Israel, and some of the US Army's original coding decisions were not transparent. Given these weaknesses, I have chosen not to use this data.[4]

Ryan Grauer and Michael Horowitz developed a new dataset to examine military effectiveness. They chose the decisive operations from all interstate wars in the twentieth century and coded each operation for which side achieved tactical victory and how well they employed forces according to Biddle's modern system, at both the operational and tactical levels. They show that forces implementing the modern system were more likely to win the eighty-three military engagements for which they had complete data.[5]

Table A.1 replicates Grauer and Horowitz' main results, then includes the *War college* and *Staff college* variables, using Models 3 and 5 from their study. The dependent variable in these models equals 1 if the side won the tactical engagement and 0 if it lost. The first set of models (based on Grauer and Horowitz' Model 3) examines military effectiveness from a monadic perspective, in other words, without taking into account a side's strength or weakness relative to the opposing side. Including data on military schools in these models does not alter Grauer and Horowitz' findings. Having a war college is correlated with tactical victory, while controlling for the variables in Grauer and Horowitz' model, but having a staff college is not correlated with tactical victory.

Grauer and Horowitz' main findings are also preserved in the results based on their model 5, which compares information between battlefield participants in each observation (in other words, a dyadic comparison). Having a war college or staff college is correlated with tactical victory, while adopting the modern system of tactics continues to be correlated with tactical victory. Grauer and Horowitz' findings about regime type also hold: consolidated democracies and consolidated autocracies are just as likely, when controlling for adoption of the reflection system (as measured by the presence of a staff or war college), to be successful on the battlefield.

Table A.1 Military Effectiveness (Probit), 1913–2003

	Grauer/Horowitz Model 3			*Grauer/Horowitz Model 5*		
	Original	*Staff College*	*War College*	*Original*	*Staff College*	*War College*
Staff college		.287 (.315)			1.993*** (.584)	
War college			.718* (.402)			4.378*** (.932)
Modern system adoption level	.497*** (.165)	.520*** (.164)	.516*** (.152)			
Comparative modern system adoption level				1.442** (.568)	1.683*** (.469)	2.919*** (.551)
Polity score	–.160 (.114)	–.152 (.112)	–.137 (.108)	–.644*** (.246)	–.880*** (.286)	–.859*** (.219)
Polity squared	.010* (.005)	.009* (.005)	.008* (.005)	.026** (.011)	.033** (.013)	.024** (.011)
Polity side B score				.347 (.236)	.478* (.257)	.813*** (.260)
Polity side B squared				–.019* (.010)	–.027** (.011)	–.043*** (.011)
Material capabilities	6.477*** (2.417)	6.392*** (2.276)	5.467*** (2.012)			
Comparative military capabilities				2.333*** (.637)	2.236** (.941)	1.806 (1.282)
Strategy (maneuver/attrition/punishment)				.333 (.275)	.878*** (.312)	1.043*** (.347)
Terrain	.538 (.872)	.558 (.888)	.540 (.888)	3.483* (2.060)	5.429*** (1.727)	8.190*** (2.023)
Troops engaged	1.35xe^{-6} (1.01xe^{-6})	1.31xe^{-6} (9.75xe^{-7})	1.32xe^{-6} (8.89xe^{-7})			

	Grauer/Horowitz Model 3			Grauer/Horowitz Model 5		
	Original	*Staff College*	*War College*	*Original*	*Staff College*	*War College*
Opponent troops engaged	$-2.16\times e^{-6}$* ($1.31\times e^{-6}$)	$-2.07\times e^{-6}$ ($1.29\times e^{-6}$)	$-2.09\times e^{-6}$* ($1.22\times e^{-6}$)			
Comparative troops engaged				.566 (1.678)	−1.405 (1.355)	−3.025* (1.675)
Comparative human capital				.002*** (.000)	.003*** (.001)	.005*** (.000)
Comparative civil-military relations				−.205 (.310)	−.307 (.543)	−.907 (.657)
Constant	−1.353 (.988)	1.562 (.996)	−1.536* (.918)	−4.644*** (1.640)	−6.799*** (1.542)	−10.013*** (1.718)
N (observations)	104	104	104	83	83	83
N (countries)	39	39	39	36	36	36
Pseudo-R^2	.398	.404	.424	.755	.799	.840
Probability>χ^2	.000	.000	.000	.000	.000	.000

Note: Robust errors clustered on country code; *=$p<.10$, **=$p<.05$, ***=$p<.01$, two-tailed.

In terms of overall model fit, the models that include the *War college* variable perform better than the comparison models from the Grauer and Horowitz study. For model 3, including the *War college* variable increases the pseudo-R^2 statistic by 2.6 percent, and for model 5 by 8.5 percent. Including information on military schools in the Grauer and Horowitz models, and in particular the presence of a war college, increases our ability to explain when military forces will achieve tactical victory.

That said, replicating Grauer and Horowitz' results constrains the analysis to the engagements for which data for all the covariates are available, but their complete dataset actually contains force employment data on 125 engagements. Table A.2 shows predicted probabilities using all 125 of Grauer and Horowitz' cases. Since the *War college* variable performed better in the analyses reported in Table A.1, and since it is a more discriminating measure of a state's military education system, I regressed it on Grauer and Horowitz' four-point *Msadopt* variable, which indicates the extent to which a military force employed the modern system in battlefield engagements. There are no control variables entered into the analysis, as the point of correlating indicators of military education with indicators of battlefield effectiveness is to establish a fundamental assumption of this study, which is that the way

Table A.2 Predicted Probabilities, Adopting the Modern System (Grauer & Horowitz)

		Probability of Adopting the Modern System	
Value	*Description*	*Without War College*	*With War College*
0	Failure to implement/minimal implementation of modern system at tactical and operational levels	.38	.20
1	Moderate implementation at one level (tactical or operational), failure to implement/minimal implementation at the other level	.22	.19
2	Moderate implementation at both tactical and operational levels, or thorough implementation at one level and no implementation at the other	.15	.16
3	Thorough implementation at both levels, or thorough implementation at one level and moderate implementation at the other	.25	.44

Note: Definitions come from Grauer & Horowitz 2016; predicted probabilities obtained using King et al.'s (2000) CLARIFY program, based on regressing the *War college* dummy (1=country has a war college at the start of a conflict) on Grauer & Horowitz' *Msadopt* variable using ordered probit (coefficient: .535, standard error: .210, *p*=.011, *n*=125).

militaries think influences the way they fight. Future research will engage this relationship more fully.

I used the CLARIFY program to predict the probability that a state with a war college will fall on a given value of Grauer and Horowitz' four-point scale (Table A.3).[6] In the 125 cases for which we have data, states with a war college are 19 percent more likely to implement the modern system of tactics (level 3 on the scale) and 18 percent less likely to not implement it (level 0). At the midrange of the scale, the distinction between having and not having a war college is not that stark, although the 18–19 point swing at the ends of the scale is substantial.

Because military effectiveness could also be defined as the ability of a military force to achieve political victory, I conduct robustness checks using data on 140 interstate war outcomes from 1945 to 2003, from the Correlates of War Project's (COW) Interstate War Dataset.[7] I examined only wars after 1945 given the unique geopolitical circumstances of the postwar era. I estimated the probability that a state with a war college (using the *War college* variable) would win an interstate war. I expect this relationship to be positive because militaries whose officers think about how to fight better in the schoolhouse should actually lead better on the battlefield, resulting ultimately in political victory.

The COW Project provides an eight-point nominal level variable (*Outcome*) where 1 means that the side in question won the war, 2 means the side lost, 3 means the result was a tie, 4 means the war transformed into another type of war (e.g., intrastate, extrastate), and so on. I transformed this variable into a dummy variable (*Winner*) that equals 1 if the COW *Outcome* variable equals 1, and 0 otherwise, which presents a high bar for the notion that states with more developed military education systems are more likely to win wars, since the *Winner* variable includes only clear-cut cases of victory.

Table A.3 Predicted Probabilities, Interstate War Outcomes, 1945–2003

Value	*Description*	*Probability* without a *War College*	*Probability* with a *War College*	*Change in Probability*
0	Country did not win an interstate war (*n*=84)	.77	.57	−.20
1	Country won an interstate war (*n*=56)	.23	.43	+.20

Note: Interwar outcome data from the Correlates of War Project; baseline probability of winning an interstate war after 1945: .301; predicted probabilities obtained using King et al.'s (2000) CLARIFY program, based on regressing the *War college* dummy (1=country has a war college at the start of a conflict) on war outcome data (dummy) using logit estimation (coefficient: .942, standard error: .359, *p*=.009, *n*=140, clusters=61).

Even given this high bar for establishing a link between military education and battlefield effectiveness, the data on interstate wars since 1945 indicate a 20 percent swing in the probability of political victory if a state has a war college. Thus, having an advanced military education system matters for both tactical performance and political victory in war.

But assessing victory in war is a tricky thing, and not always connected directly to a military's performance, so I also examined another element of battlefield effectiveness: fatality levels in militarized interstate disputes (MIDs). Even though many of these disputes are not out-and-out shooting affairs, like fishing disputes or border standoffs, I argue that the system of military reflection, if it is truly systemic, should have an effect on performance even if the use of force is merely threatened. Whereas the Grauer and Horowitz and COW interstate war data measure tactical and political outcomes, the level of fatalities during a militarized dispute, even at the lowest levels of violence, measures the efficiency of a military force in achieving these objectives.

The results are presented in Table A.4, and examine all MIDs from 1816 to 2010. MIDs include not only interstate wars, but also the use or threat of military force between a pair of states. The data, also drawn from the COW Project,[8] provides a *Fatality* variable, which lists the level of combat fatalities that states suffer in a dispute, ranging on a seven-point scale from 0 (no fatalities) to 6 (over 999 fatalities). By estimating an ordered probit model using the *War college* variable, I derive the predicted probability that states will fall into each of these fatality categories. As Table A.4 shows, having a war college creates a stark difference at the high and low ends of the scale. States with a war college are 5.9 percent more likely to suffer no fatalities during a MID, and 1.7 percent less likely to suffer 1,000 or more fatalities.

Since this result considers all MIDs in the modern era, it further strengthens the robustness of the link between military education and battlefield

Table A.4 Probability of Fatalities during Militarized Interstate Disputes, 1816–2010

Value	*Description*	*Without War College*	*With War College*
0	No fatalities	.809	.868
1	1–25 deaths	.095	.071
2	26–100 deaths	.028	.019
3	101–250 deaths	.011	.007
4	251–500 deaths	.008	.006
5	501–999 deaths	.007	.004
6	Over 999 deaths	.042	.025

Note: MID data from the Correlates of War Project; predicted probabilities obtained using King et al.'s (2000) CLARIFY program, based on regressing the *War college* dummy (1=country has a war college at the start of a conflict) on MIDB *Fatality* variable using ordered probit (coefficient: –.242, standard error: .081, *p*=.003, *n*=5,321, clusters: 156).

effectiveness. Having a war college corresponds with suffering fewer casualties during disputes, winning interstate wars, winning battles, and implementing the modern system of tactics. This supports the fundamental assumption of this study that adopting the system of military reflection contributes to battlefield success.

ADOPTING THE SYSTEM OF MILITARY REFLECTION

If military education and adopting the system of military reflection contribute to battlefield effectiveness, then what leads states to reform military education and adopt the system of military reflection in the first place? Chapter 1 of this study examined five hypotheses that could answer this question:

Regime Type Hypothesis: Democratic states are more likely to reform military education.

International Security Hypothesis: States with greater security concerns are more likely to reform military education.

Cultural Distance Hypothesis: States with more cultural distance to cross to implement reforms are less likely to reform military education.

Wealth Hypothesis: Wealthier states are more likely to reform military education.

Human Capital Hypothesis: States with a higher level of human capital are more likely to reform military education.

Whereas the analysis linking military education and battlefield effectiveness treated military education as the explanatory factor, the analysis of these five hypotheses treats reform in military education and the adoption of the system of military reflection as the factor to be explained. I use four indicators to assess when and how countries adopt the system of military reflection: the establishment of a staff college, the establishment of a war college, the number of military periodicals, and military expenditures per soldier. Other scholars have used military expenditures per soldier to measure the quality of military forces,[9] and data for this measure come from the COW Project's National Military Capabilities dataset.[10] For the analysis below, the military expenditures per soldier data is transformed using a logarithmic function, and I report ordinary least squares regression results for this indicator as a result.

The other three indicators of adoption are original. In previous research, I describe the data collection for military periodicals, but the reason that military periodicals indicate adoption of the system of military reflection is that they provide a forum for the cultivation and dissemination of knowledge specific to the military profession. They are a mechanism of institutional introspection, and they help define a military's role in a country's civil-military relations. Perhaps most importantly, they also constitute a venue for military officers to engage in critical analysis of military operations, both past and present.

The *military periodicals* variable is a count of the number of periodicals dealing with military affairs that are published in a country in a given year, as indexed by Ulrich's Periodicals Directory and WorldCat.[11] Data extend only to 2005, but there is a significant amount of variation across the world, with a substantial number of countries publishing no military periodicals, and most publishing less than five. Two outliers, the United States and the United Kingdom, published a substantial majority of military periodicals in the post-World War II era, which is why it is important to use alternative indicators of adopting the system of military reflection, and to cluster standard errors on country code. Given the large number of observations with no military periodicals, I used negative binomial regression to analyze the military periodicals data.

The last two indicators of adopting the system of military reflection are *War college* and *Staff college*. As previously described, these two dummy variables indicate whether a state has at least one war or staff college in a given year. A staff college is a school to prepare military officers as they transition from tactical command to positions of strategic concern, where they will be responsible for organizing military operations in order to achieve strategic objectives. A war college is a school to prepare military officers and future civilian leaders in strategic decision-making, or in how to organize all elements of national power to preserve national interests and achieve national goals. The presence of staff and war colleges suggest adoption of the system of military reflection because, similar to military periodicals, they provide a venue for critical analysis, introspection, and defining a country's civil-military relations.

The vast majority of states today boast staff colleges, but there is wide variation in the timing of when these schools were founded. In 2015, only sixty-three states had a war college. Given the binary nature of these indicators, I analyze these data using logit estimation, although I begin the analysis with Cox survival models. This allows me to examine separately the reasons why states start reforms (establishing military schools, for which I use the Cox models) and why they maintain these reforms in the long term, for which I use the logit models. Both questions are relevant to understanding how states adopt the system of military reflection.

Tables A.5, A.6, and A.8 use a number of variables to test the five hypotheses above. To test the Wealth Hypothesis, I use data on gross domestic product per capita, in constant 1990 US dollars (*GDP per capita*). Data are log-transformed, and they come from the Maddison Project. These data are the most widely used to examine wealth in years prior to 1945, and coverage is fairly comprehensive.[12]

To test the Human Capital Hypothesis, I use two variables from the World Bank's EdStats dataset: *Grade 1 intake* and *Tertiary education expectancy*. *Grade 1 intake* refers to the proportion of children in a population that enter first grade at the appropriate age, and *Tertiary education expectancy* refers to the number of years of tertiary (college) education an individual in a population is expected to attain in his or her lifetime. These two indicators assess human capital at the beginning and end of an education system, they have considerable coverage across countries, and a factor analysis suggested that they are different enough from one another to be used in the same models. However, these data are only available from 1970, so models using these indicators are limited to the post-1970 period.[13]

I use five indicators to test the International Security Hypothesis, not only because realist interpretations predominate in international relations scholarship, but also because previous research has shown that security concerns to have an impact on how states organize military power.[14] *Interstate rivalry* is based on the work of James Klein, Gary Goertz, and Paul Diehl, and equals 1 if a state was party to at least one interstate rivalry by their definition, and 0 otherwise. In robustness checks (not reported), I ran models using two competing indicators of interstate rivalry, but this did not change the substantive interpretation of the result. Since the Klein et al. data seems to be used more frequently in the literature; I report models using those data here.[15]

The other four indicators of international security concerns are based on data from the COW Project. *Recent war participation* equals 1 if a state is currently or has in the last five years been involved in an interstate war, and is included because recent battlefield experience, whether positive or negative, makes it more likely that states will have military lessons to learn, and because, according to neorealist logic, competition in the state system is the driver of socialization.

The *recent great power war* variable equals 1 if a war between two or more great powers is either ongoing or ended within the last five years, and 0 otherwise. This is a systemic dummy variable, as opposed to being specific to a country, and it is included because, anecdotally, states have established military schools and began military reform on the heels of great power wars, even when they were not involved. In South America, for example, German military advisors were in high demand after the Wars of German Unification and the Franco-Prussian War, and a flurry of military reform activity occurred around the globe following World War II.[16]

Great power equals 1 if the COW Project considers a state a major power in the year in question, and 0 otherwise. It is included because of the markedly different security concerns experienced by great powers as compared to other states, concerns that could contribute to the decision to adopt and maintain the system of military reflection.

The final indicator of international security concerns is *Material capabilities*, which is the Combined Index of National Capabilities (CINC) score from the COW Project's National Material Capabilities dataset. This index includes information on military expenditures and force size, manufacturing and industrial capacity, and population (total and urban). I expect it to assess any aspects of the international security environment not captured by the other four indicators.[17]

To test the Regime Type Hypothesis, I use Carles Boix, Michael Miller, and Sebastian Rosato's dichotomous measure of democracy (*Democracy*). This indicator is more elegant than the well-known Polity IV Project's data, which places countries on two ten-point scales of democracy and autocracy, and which is often transformed into a dichotomous variable anyway. Boix, Miller, and Rosato's measure was coded specifically as a dichotomous variable, and focuses on the essential elements of democracy: contestation and participation. This standard of discrimination might present a lower bar for what constitutes a democracy, which constitutes an easier test for the Regime Type Hypothesis and a harder test for adoption theory. At the same time, models using both indicators were substantively equivalent, so the models reported here use the easier-to-understand Boix et al. measure of democracy.[18]

To test the Cultural Distance Hypothesis, I use information on a state's relationship with its former colonial ruler, since previous research has shown that being a former British colony is strongly associated with using volunteer military recruitment as opposed to conscription,[19] and a similar relationship might hold for other military institutions. Anecdotal observations indicate that former colonies often pattern their post-independence military education systems on those of the former colonial ruler. *Colonial ruler model* refers to the former colonial ruler's behavior on an indicator of adopting the system of military reflection. For example, in models of whether a state has established a war college or not, this variable equals the number of years that have passed since the former colonial ruler established a war college. In models of military periodicals and military expenditures per soldier, this variable refers to the values of the former colonial ruler, in that year, for the relevant indicator of adopting the system of military reflection. The *Violent independence* variable equals 1 if a state's independence was violent, and 0 otherwise. Data for these two variables come from Paul Hensel's ICOW Colonial History Dataset.[20] An interaction term for these two variables means that, in models including all three variables, coefficients for the interaction term assess the impact of the colonial ruler's behavior in cases where a state's independence was violent, and coefficients for the *Colonial ruler model* assess the impact of the colonial ruler's behavior only in cases where independence was not violent.

The models include one control variable, *Time since independence*, which has had an effect in previous models of military professionalization.[21] The longer a state maintains its independence, the more likely it is to develop its capacity, which could include adopting the system of military reflection.

ANALYSIS

Table A.5 shows results for Cox survival models of the propensity of states to establish staff and war colleges.[22] The wealth models examine the period after 1816, and the human capital models that after 1970. The human capital models do not include the *recent war participation*, *recent great power war*, and *great power* variables due to data coverage. There were no great power wars in the 1970–2011 period, and no great powers or states that had recently been involved in an interstate war that created their first war college in this period, so the models could not provide estimates for those variables.

These Cox survival models are telling for the variables that are not statistically significant. *GDP per capita* does not appear to be associated with the timing of founding staff and war colleges, and only the *Tertiary education expectancy* is significant, and that only for the establishment of staff colleges. It appears, then, that neither the Wealth Hypothesis nor the Human Capital Hypothesis is supported when it comes to founding military schools.

The International Security Hypothesis fares slightly better. In the post-1816 period, states were systematically more likely to found war colleges on the heels of great power wars, and to establish war or staff colleges if they had higher material capabilities relative to the rest of the world. In addition, being involved in an interstate rivalry was correlated with the establishment of staff colleges.

The Regime Type Hypothesis receives no support in these Cox survival models, but the Cultural Distance Hypothesis garners qualified support. Because the *violent independence* and *colonial ruler model* variables are included along with an interaction term, the coefficients for the *colonial ruler model* reflect only to the number of years since a former colonial ruler established a war college in the event of peaceful independence. Given this, the two models of the post-1816 period suggest that the connection with the former colonial ruler influences the establishment of a war college when independence was peaceful. At the same time, this effect does not seem to hold in the post-1970 period, perhaps because independence in later years means

Table A.5 Establishing Military Schools (Cox Survival Models), 1816–2011

	Wealth Models (1816–2011)		*Human Capital Models (1970–2011)*	
	War College	*Staff College*	*War College*	*Staff College*
GDP per capita (log)	.789 (.138)	1.097 (.128)		
Grade 1 intake			1.009 (.012)	.996 (.012)
Tertiary education expectancy			1.731 (.751)	2.321* (.772)
Interstate rivalry (dummy)	1.577 (.532)	1.774** (.480)	.581 (.305)	1.438 (.597)
Recent war participation (dummy)	1.161 (.540)	1.231 (.428)		
Recent great power war (systemic dummy)	2.782** (1.288)	1.745 (.592)		
Great power (dummy)	.693 (.722)	.823 (.595)		
Material capabilities	$6.67\times e^{10}$*** ($5.68\times e^{11}$)	$3.71\times e^{6}$*** ($1.54\times e^{7}$)	10.748 ($5.428\times e^{2}$)	$9.768\times e^{2}$ ($1.033\times e^{4}$)
Democracy (dummy)	1.686 (.656)	.916 (.270)	.707 (.560)	.921 (.488)
Colonial ruler model	.984** (.008)	.986** (.006)	.980 (.012)	.994 (.009)
Violent independence (dummy)	.961 (.391)	1.351 (.419)	1.203 (.788)	2.430 (1.597)
Colonial ruler model × Violent ind.	1.011 (.010)	1.007 (.007)	1.017 (.013)	.996 (.012)
Time since independence	1.010** (.004)	.999 (.003)	1.002 (.006)	.995 (.004)
N (observations)	8,408	6,002	2,269	1,719
N (countries)	154	150	124	104
Probability>χ^2	.000	.000	.000	.000

Note: Robust errors clustered on country code; *=p<.10, **=p<.05, ***=p<.01, two-tailed.

something different than it did earlier, and it necessitates a clean break with the colonial past. As for the *democracy* variable, being a democracy is not associated with when states establish staff or war colleges.

In sum, the only hypothesis to receive somewhat consistent support in these models is the International Security Hypothesis. To explain the timing of establishing staff and war colleges, the factors that matter most seem to be whether a state is engaged in a rivalry, the extent of its material capabilities, and whether a great power war has recently occurred in the international system. Levels of economic wealth and human capital play little to no role in the timing of establishing military schools.

But establishing military schools is not the same as maintaining the system of military reflection over a long period. Table A.6 presents results that examine this related question. These models include the same variables as those reported in Table A.5, although instead of examining why states establish military schools, the data is set up to examine the probability that a state will have a staff college or war college in any given year. In addition, the human capital variables were not included because they would have constrained cases to after 1970. These models also expand beyond military schools to examine the other two indicators of adopting the system of military reflection: the number of military periodicals and military expenditures per soldier. The *material capabilities* variable was dropped from the model of military expenditures per soldier, since the latter is a component of the former.

In contrast to the models meant to explain the initiation of reforms to adopt the system of military reflection, the Wealth and Cultural Distance Hypotheses receive consistent support throughout the 1816–2011 period. Only one of the factors linked with the International Security Hypothesis, *material capabilities*, is consistently associated with indicators of the reflection system, and in the anticipated direction. Being a great power, experiencing an interstate rivalry, being involved in a recent interstate war, and witnessing a great power war in the state system are associated with only two or three of the four indicators of the system of military reflection. As for the Regime Type Hypothesis, being a democracy is associated with establishing a war college and publishing more military periodicals, but democracies are less likely to have staff colleges and there seems to be no relationship with martial quality, as measured by military expenditures per soldier.

The level of economic wealth and the former colonial ruler's behavior in cases of peaceful independence, though, are associated with maintaining staff and war colleges, with publishing more military periodicals, and with spending more per soldier. Thus, the Wealth and Cultural Distance Hypotheses are supported by these models.

Table A.7 illustrates the importance of wealth relative to other factors that might also correlate with maintaining the system of military reflection.

Table A.6 Maintaining the System of Military Reflection, 1816–2011

	War College (logit)	*Staff College (logit)*	*Military Periodicals (Negative Binomial)*	*Military Expenditure per Soldier (log) (OLS)*
GDP per capita (log)	.165*** (.050)	.401*** (.031)	.674*** (.134)	1.062*** (.100)
Interstate rivalry (dummy)	1.166*** (.080)	.336*** (.052)	.212 (.144)	.234* (.134)
Great power (dummy)	.330*** (.116)	–.334*** (.115)	.369 (.378)	–1.117*** (.427)
Material capabilities	4.598** (.718)	11.786*** (.916)	7.952** (3.122)	
Recent war participation (dummy)	.622*** (.094)	.517*** (.072)	.291*** (.082)	.151 (.112)
Recent great power war (systemic dummy)	–1.469*** (.106)	.022 (.061)	–.008 (.100)	–.832*** (.117)
Democracy (dummy)	.824*** (.090)	–.155** (.061)	.546** (.239)	.096 (.136)
Colonial ruler model	.006*** (.002)	.007*** (.001)	.002*** (.000)	.324*** (.035)
Violent independence (dummy)	.193** (.087)	.554*** (.092)	–.252 (.207)	–.553 (.468)
Colonial ruler model × violent independence	–.009*** (.003)	–.001* (.001)	.001 (.001)	–.010 (.063)
Time since independence	.014*** (.004)	.002 (.003)	.011 (.009)	–.008 (.006)
Constant	–5.427*** (.390)	–6.401*** (.253)	–6.412*** (1.017)	–2.570*** (.761)
N (observations)	9,854	9,854	9,710	7,136
N (countries)	156	156	156	156
Pseudo-R^2	.324	.195	.225	
Probability>χ^2	.000	.000	.000	
R^2				.572
Probability>F				.000

Note: Robust errors (clustered on country code for *military periodicals* and *military expenditures per soldier* models); *=p<.10, **=p<.05, ***=p<.01, two-tailed; all models include time, time-squared, and time-cubed variables (coefficients not reported).

It displays predicted probabilities based on the war college model in Table A.6. In simulations based on this model, moving from one standard deviation below to one standard deviation above the mean on *GDP per capita* meant a 1.3 percent increase in the probability that a state would have a war college. A one standard deviation increase in material capabilities increased the probability of having a war college by 1.2 percent. For those states that experienced peaceful independence, going from the former colonial ruler not having established a war college (the modal category: 0) to the mean for this variable (the former colonial ruler having had a war college for about twenty-five years) translated into a 0.9 percent increase in the probability that a state would have a war college, although this same transition for states that experienced violent independence actually *decreased* the probability of having a war college by 0.4 percent. This may mean that cultural affinity is dependent on whether a state's independence was violent or not. So, while the Wealth Hypothesis is supported across a variety of indicators of adopting the system of military reflection, there is reason to believe that, under certain conditions, the International Security or Cultural Distance Hypotheses could also have an impact on maintaining the system of military reflection.

That said, the evidence presented thus far does not examine the Human Capital Hypothesis. Table A.8 replicates the models from Table A.6 using indicators of human capital accumulation instead of wealth accumulation.

Table A.7 Predicted Probability of Having a War College

If...	*...is set to...*	*...then the probability of having a war college is:*	*Change in Probability*
GDP per capita (log)	mean	.076	+.013
GDP per capita (log)	One SD *above* mean	.089	
Material capabilities (CINC score)	mean	.076	+.012
Material capabilities (CINC score)	One SD *above* mean	.088	
Colonial ruler model (peaceful independence)	0 (modal category)	.067	+.009
Colonial ruler model (peaceful independence)	mean	.076	
Colonial ruler model × Violent independence	0 (modal category)	.080	−.004
Colonial ruler model × Violent independence	mean	.076	

Note: All other variables set to their means; baseline probability (all variables set to their means): .084; probabilities generated using King et al.'s (2000) CLARIFY program.

Table A.8 Human Capital and the System of Military Reflection, 1970–2011

	War College (logit)	*Staff College (logit)*	*Military Periodicals (Negative Binomial)*	*Military Expenditure per Soldier (log) (OLS)*
Grade 1 Intake	.016*** (.004)	.007*** (.002)	.017*** (.004)	.007* (.003)
Tertiary education expectation	.555*** (.101)	.696*** (.107)	.577*** (.144)	.688*** (.097)
Interstate rivalry (dummy)	.230* (.132)	.085 (.101)	.096 (.191)	–.108 (.138)
Great power (dummy)	.884 (.635)	2.028*** (.761)	.562 (.470)	.817*** (.295)
Recent war participation (dummy)	–.109 (.218)	.901*** (.168)	.208 (.134)	.065 (.144)
Material capabilities	148.928*** (11.241)	39.046*** (8.941)	38.334** (15.597)	
Democracy (dummy)	–.048 (.167)	–.575*** (.120)	–.151 (.268)	.216 (.167)
Colonial ruler model	–.042*** (.004)	.005*** (.001)	.001* (.000)	.081* (.047)
Violent independence (dummy)	–1.077*** (.166)	.377 (.237)	–.108 (.265)	.093 (.537)
Colonial ruler model × violent independence	.026*** (.005)	.002 (.002)	–.001 (.001)	–.041* (.065)
Time since independence	–.004 (.006)	–.009** (.005)	.016** (.008)	.001 (.003)
Constant	–2.950*** (.441)	–5.392*** (.443)	–3.277*** (.523)	7.282*** (.516)
N (observations)	2,905	2,905	2,880	2,315
N (countries)	151	151	151	148
Pseudo-R^2	.495	.306	.235	
Probability>χ^2	.000	.000	.000	
R^2				.387
Probability>F				.000

Note: Robust errors (clustered on country code for *military periodicals* and *military expenditures per soldier* models); *p<.10, **p<.05, ***p<.01, two-tailed; all models include time, time-squared, and time-cubed variables (coefficients not reported).

The results suggest support for the Human Capital Hypothesis, at least in years since 1970. Both human capital indicators, *grade 1 intake* and *tertiary education expectation*, are consistently associated with all four indicators of the system of military reflection, and they are in the anticipated direction. The only other indicator that exhibits the same behavior is *material capabilities*. The other international security measures show inconsistent associations with the reflection system indicators. Taken together, this suggests weaker support for the International Security, Regime Type, and Cultural Distance Hypotheses than for the Human Capital Hypothesis. Having access to human capital, then, matters for maintaining the system of military reflection.

ADOPTION THEORY

This statistical evidence suggests two stages in the process of adopting the system of military reflection. The Wealth, Human Capital, Regime Type, and Cultural Distance Hypotheses receive little support in these data at the initial reform stage, when states first decide to adopt the system of military reflection by founding staff and war colleges. The International Security Hypothesis performs best at this stage, suggesting that international security concerns catalyze the motivation to reform.

But economic modernization makes this reform a reality over the longer term. In statistical models examining the maintenance of the system of military reflection, the Wealth and Human Capital Hypotheses received consistent support over a variety of indicators and time periods. There may be circumstances, as for the maintenance of war colleges, when the International Security and Cultural Distance Hypotheses demonstrate greater relative impact, but only different analytical methods (such as case studies) could tease out these relationships. Maintaining the system of military reflection seems to be associated with states' material capabilities, but this is only an indirect measure of the international security environment. At this juncture, the statistical evidence suggests that security concerns precipitate reforms while wealth and human capital institutionalize them.

As chapter 1 describes, adoption theory suggests how these two forces—political will born of security concerns and wealth and human capital born of economic modernization—interact, generating four different pathways to adopting the system of military reflection: radical, front-loaded, fitful, or steady. The case studies in chapters 2 through 4 examine these pathways more closely. The statistical evidence presented here, however, suggests that states that aspire to military excellence cannot rely only on security concerns to motivate adoption of the system of military reflection. Over the long term, they also need modern economies.

NOTES

1. Toronto 2017.

2. Some educational institutions were called a war college prior to 1874, such as Berlin's Kriegsakademie, but no schools prepared officers and civilian officials with this national strategic insight and context until 1874.

3. Biddle and Long 2004; Ramsay 2008; Beckley 2010; Pilster and Böhmelt 2011.

4. I am also agnostic as to how military effectiveness is defined. Some scholars, like Biddle and Long 2004, examine loss-exchange ratios, or the ratio of friendly-to-enemy casualties, which depends on battle outcomes or political victory. While I am most convinced by Grauer and Horowitz' approach of examining decisive operations, I report a number of robustness checks that go beyond this, to show that the system of military reflection does indeed have systemic effects on how a military performs, in all aspects of warfare.

5. Grauer and Horowitz 2012.

6. King et al. 2000; Tomz et al. 2003.

7. Sarkees and Wayman 2010.

8. Version 4; Palmer et al. 2015.

9. Kollias 1996; Powell 2012; Kentor and Kick 2015.

10. Version 4.0; Singer et al. 1972; Singer 1987.

11. FirstSearch 2005; Ulrich's Periodicals Directory 2005; Toronto 2017.

12. Maddison Project 2013; Bolt and van Zanden 2014.

13. World Bank 2016. The post-1970 time period, combined with a lack of data coverage across countries, also forced me to drop three of the five international security variables from the human capital models: *recent war participation*, *recent great power war*, and *great power*.

14. Waltz 1979; Posen 1993; Resende-Santos 1996; Asal et al. 2015; Toronto 2017; Cohn and Toronto 2016.

15. Klein et al. 2006. See also Colaresi and Thompson 2002; Hewitt 2005.

16. Nunn 1983; Ralston 1990; Resende-Santos 1996.

17. Singer et al. 1972; Singer 1987; Sarkees and Wayman 2010; COW Project 2011; Palmer et al. 2015.

18. Boix et al. 2012; CSP 2016.

19. Asal et al. 2015; Cohn and Toronto 2016.

20. Hensel 2014.

21. Toronto 2017.

22. Box-Steffensmeier and Jones 2004.

Bibliography

Abdulla, Abdulkhaleq. "Why the UAE's Regional Activism?" *Gulf News* (January 12, 2015). http://gulfnews.com/opinion/thinkers/why-the-uae-s-regional-activism-1.1440114 (accessed November 14, 2016).

Abu-Bader, Suleiman, and Aamer S. Abu-Qarn. "Government Expenditures, Military Spending and Economic Growth: Causality Evidence from Egypt, Israel, and Syria." *Journal of Policy Modeling* 25, no. 6/7 (2003): 567–83.

Abu Dhabi University. "Military Programs." http://www.adu.ac.ae/mp/ (accessed November 14, 2016).

Aftalion, Florin. *The French Revolution: An Economic Interpretation*. Cambridge: Cambridge University Press, 1990.

Agüero, Felipe. "The Military and the Limits to Democratization in South America." In *Issues in Democratic Consolidation: The New South American Democracies in Comparative Perspective*, eds. Scott Mainwaring, Guillermo O'Donnell, and J. Samuel Valenzuela, 153–98. South Bend, IN: Notre Dame University Press, 1992.

Ahmad, Feroz. *Turkey: The Quest for Identity*. Oxford: Oneworld, 2003.

Akmeşe, Handan Nezir. *The Birth of Modern Turkey: The Ottoman Military and the March to World War I*. London: I.B. Tauris, 2005.

Aktan, R. "Agricultural Policy of Turkey." In *The Economic History of the Middle East, 1800–1914: A Book of Readings*, ed. Charles Issawi, 107–13. Chicago, IL: University of Chicago Press, 1966.

Alagappa, Muthiah, ed. "Investigating and Explaining Change: An Analytical Framework." In *Coercion and Governance: The Declining Political Role of the Military in Asia*, 29–68. Stanford, CA: Stanford University Press, 2001.

Albrecht, Holger, and Dina Bishara. "Back on Horseback: The Military and Political Transformation in Egypt." *Middle Eastern Law & Governance* 3, no. 1/2 (2011): 13–23.

Altinay, Ayse Gül. *The Myth of the Military-Nation: Militarism, Gender, and Education in Turkey*. New York, NY: Palgrave Macmillan, 2004.

American Anthropological Association (AAA). "Statement on Ethics: Principles of Professional Responsibility." *AAA Ethics Blog* (November 1, 2012). http://ethics.americananthro.org/category/statement/ (accessed November 14, 2016).

Anderson, Robert David. *France 1870–1914: Politics and Society*. London: Routledge and Kegan Paul, 1977.

Anonymous ('A. Puzzled Prof'). "The Collapse of NDU: A View from Inside." *Best Defense* (June 19, 2012). http://foreignpolicy.com/2012/06/19/the-collapse-of-ndu-a-view-from-inside/ (accessed March 18, 2015).

Anonymous (A French Soldier). "A French Soldier's View of US Soldiers in Afghanistan." *Warrior Lodge* (December 15, 2014). http://www.warriorlodge.com/blogs/news/16298760-a-french-soldiers-view-of-us-soldiers-in-afghanistan (accessed April 12, 2015).

Asal, Victor, Justin Conrad, and Nathan Toronto. "I Want You! The Determinants of Military Conscription." *Journal of Conflict Resolution* 61, no. 7 (August 2017): 1456–81.

Azzam, Maha. "Egypt's Military Council and the Transition to Democracy." http://www.chathamhouse.org/sites/default/files/public/Research/Middle%20East/bp0512_azzam.pdf (accessed August 19, 2012). London: Chatham House, 2012.

al-Bahi, Ibrahim, Maher Maqlad, and Amani Magid. "إحالة طنطاوي والإانان للتكعد: السيسي وزيرا للدفاع: مرسي يلغي الدستور المكمل ويعين محمود مكي نائبا لرئيس الجمهورية" (Tantawi and Anan Sent to Retirement: Sisi Minister of Defense: Mursi Abrogates Constitutional Proclamation and Appoints Mahmoud Mekki Vice President of the Republic. *Al-Ahram* (August 12, 2012). http://www.ahram.org.eg/Inner.aspx?ContentID=165375&typeid=25 (accessed August 19, 2012).

Barany, Zoltan. "The Role of the Military." *Journal of Democracy* 22, no. 4 (2011): 24–35.

Barnett, Michael N. *Confronting the Costs of War: Military Power, State, and Society in Egypt and Israel*. Princeton, NJ: Princeton University Press, 1992.

———. *Dialogues in Arab Politics: Negotiations in Regional Order*. New York, NY: Columbia University Press, 1998.

Battles, Michael. 2017. "U.S., Egypt kick off Exercise Bright Star 2017." US Embassy & Consulate in Egypt (September 10, 2017). https://eg.usembassy.gov/u-s-egypt-kick-off-exercise-bright-star-2017/ (accessed October 5, 2017).

Beattie, Peter M. *The Tribute of Blood: Army, Honor, Race, and Nation in Brazil, 1864–1945*. Durham, NC: Duke University Press, 2001.

Beck, Hermann. "State and Society in Pre-March Prussia: The Weavers' Uprising, the Bureaucracy, and the Association for the Welfare of Workers." *Central European History* 25, no. 3 (1992): 303–31.

Beckley, Michael. "Economic Development and Military Effectiveness." *Journal of Strategic Studies* 33, no. 1 (2010): 43–79.

Be'eri, Eliezer. *Army Officers in Arab Politics and Society*. London: Praeger, 1969.

Bell, Jennifer. "Mohammed bin Zayed Opens New Defence College in Abu Dhabi." *National* (December 4, 2013). http://www.thenational.ae/uae/government/mohammed-bin-zayed-opens-new-defence-college-in-abu-dhabi (accessed November 13, 2016).

Bennett, Andrew, and Jeffrey T. Checkel. *Process Tracing: From Metaphor to Analytic Tool*. Cambridge: Cambridge University Press, 2015.

Berger, Peter L., and Thomas Luckmann. *The Social Construction of Reality: A Treatise in the Sociology of Knowledge*. London: Penguin UK, 1991.

Berkes, Niyazi. *Development of Secularism in Turkey*. Montreal: McGill University Press, 1965.

Biddle, Stephen, and Stephen Long. "Democracy and Military Effectiveness: A Deeper Look." *Journal of Conflict Resolution* 48, no. 4 (2004): 525–46.

Biddle, Stephen. *Military Power: Explaining Victory and Defeat in Modern Battle*. Princeton, NJ: Princeton University Press, 2006.

Birand, Mehmet Ali. *Shirts of Steel: An Anatomy of the Turkish Armed Forces*. New York, NY: St. Martin's Press, 1991.

Blagden, David. "Induction and Deduction in International Relations: Squaring the Circle Between Theory and Evidence." *International Studies Review* 18, no. 2 (2016): 195–213.

Boix, Carles, Michael Miller, and Sebastian Rosato. "A Complete Data Set of Political Regimes, 1800–2007." *Comparative Political Studies* 46, no. 12 (2013): 1523–54.

Bolt, Jutta, and Jan Luiten van Zanden. "The Maddison Project: Collaborative Research on Historical National Accounts." *Economic History Review* 67, no. 3 (2014): 627–51.

Bond, Kevin M. "Are We Professionals?" *Joint Force Quarterly* 58 (3rd Quarter 2010): 62–68.

Box-Steffensmeier, Janet M., and Bradford S. Jones. *Event History Modeling: A Guide for Social Scientists*. Cambridge: Cambridge University Press, 2004.

Brogan, Denis William. *The Development of Modern France, 1870–1939*, rev. ed. Gloucester, MA: Peter Smith, 1970.

Brooks, Risa A. "Making Military Might: Why Do States Fail and Succeed? A Review Essay." *International Security* 28, no. 2 (2003): 149–91.

Brown, Cameron S., Christopher J. Fariss, and R. Blake McMahon. "Recouping after Coup-Proofing: Compromised Military Effectiveness and Strategic Substitution." *International Interactions* 42, no. 1 (January 2016): 1–30.

Caforio, Giuseppe. "The Military Profession: Theories of Change." *Armed Forces & Society* 15, no. 1 (1988): 55–69.

Caldwell, William B., and Nathan K. Finney. "Security, Capacity, and Literacy." *Military Review* (January–February 2011): 23–27.

Campbell, Kirk S. "Civil-Military Relations and Political Liberalization: A Comparative Study of the Military's Corporateness and Political Values in Egypt, Syria, Turkey, and Pakistan." Ph.D. diss., George Washington University, 2009. http://gradworks.umi.com/3330856.pdf (accessed August 19, 2012).

Çelik, Zafer, and Bekir S. Gür. "Turkey's Education Policy During the AK Party Era (2002–2013)." *Insight Turkey* 15, no. 4 (2013): 151–76.

Center for Systemic Peace (CSP). "INSCR Data Page. Polity IV Annual Time-Series, 1800–2015." http://www.systemicpeace.org/inscrdata.html (accessed September 15, 2016).

Chandrasekaran, Rajiv. "In the UAE, the United States Has a Quiet, Potent Ally Nicknamed 'Little Sparta'." *Washington Post* (November 9, 2014). https://www.washingtonpost.com/world/national-security/in-the-uae-the-united-states-has-a-quiet-potent-ally-nicknamed-little-sparta/2014/11/08/3fc6a50c-643a-11e4-836c-83bc4f26eb67_story.html (accessed October 9, 2016).

Cilliler, Yavuz. "Popular Determinant on Civil-Military Relations in Turkey." *Arab Studies Quarterly* 38, no. 2 (2016): 500–20.

Cizre, Umit. "Disentangling the Threads of Civil-Military Relations in Turkey: Promises and Perils." *Mediterranean Quarterly* 22, no. 2 (2011): 57–75.

Clausewitz, Carl von. *On War*. Princeton, NJ: Princeton University Press, 1976 [1831].

Clemente, Steven E. *For King and Kaiser! The Making of the Prussian Army Officer, 1860–1914*. New York, NY: Greenwood Press, 1992.

Cohen, Eliot A. *Citizens and Soldiers: The Dilemmas of Military Service*. Ithaca, NY: Cornell University Press, 1985.

———. "Civil-Military Relations." *Orbis* 41, no. 2 (1997): 177–86.

Cohn, Lindsay P., and Nathan W. Toronto. "Markets and Manpower: The Political Economy of Compulsory Military Service." *Armed Forces & Society* 43, no. 3 (July 2017): 436–58.

Colaresi, Michael, and William R. Thompson. "Strategic Rivalries, Protracted Conflict, and Crisis Escalation." *Journal of Peace Research* 39, no. 3 (May 2002): 263–87.

Cole, Juan. "Egypt's New Left versus the Military Junta." *Social Research* 79, no. 2 (2012): 487–510.

Connelly, Owen. *Blundering to Glory: Napoleon's Military Campaigns*. Lanham, MD: Rowman & Littlefield Publishers, 2006.

Cook, Steven A. *The Unspoken Power: Civil-Military Relations and the Prospects for Reform*. Washington, DC: Saban Center for Middle East Policy at the Brookings Institution (September 1, 2004). https://www.brookings.edu/research/the-unspoken-power-civil-military-relations-and-the-prospects-for-reform/ (accessed November 16, 2016).

———. *Ruling but Not Governing: The Military and Political Development in Egypt, Algeria, and Turkey*. Baltimore, MD: Johns Hopkins University Press, 2010.

———. *The Struggle for Egypt: From Nasser to Tahrir Square*. New York, NY: Oxford University Press, 2012.

Corum, James S. *The Roots of Blitzkrieg: Hans von Seeckt and German Military Reform*. Lawrence, KS: University Press of Kansas, 1992.

———. "A Comprehensive Approach to Change: Reform in the German Army in the Interwar Period." In *Challenge of Change: Military Institutions and New Realities, 1918–1941*, eds. Harold R. Winton and David R. Mets, 35–73. Lincoln, NE: University of Nebraska Press, 2000.

Craig, Gordon A. *The End of Prussia*. Madison, WI: Univ of Wisconsin Press, 1984.

Criss, Nur Bilge. "U.S. Forces in Turkey." In *U.S. Military Forces in Europe: The Early Years, 1945–1970*, eds. Simon W. Duke and Wolfgang Krieger. Boulder, CO: Westview Press, 1993.

Cropsey, Seth, and Arthur Milikh. "Democracy in Egypt: Applying the Tocqueville Standard." *World Affairs* 174, no. 1 (May/June 2011): 51–58.

Demarest, Geoff. *Winning Insurgent War: Back to Basics*. Ft. Leavenworth, KS: Foreign Military Studies Office, 2011.

Dempsey, Martin. "Joint Education White Paper." Office of the Joint Chiefs of Staff (July 16, 2012). www.dtic.mil/doctrine/concepts/white_papers/cjcs_wp_education.pdf (accessed March 18, 2015).

Desch, Michael C. *Civilian Control of the Military*. Baltimore, MD: Johns Hopkins University Press, 1999.

———. "Democracy and Victory: Why Regime Type Hardly Matters." *International Security* 27, no. 2 (2002): 5–47.

———. *Power and Military Effectiveness: The Fallacy of Democratic Triumphalism*. Baltimore, MD: Johns Hopkins University Press, 2010.

Dreazen, Yochi J. "After Mubarak, Will Egypt Get Fooled Again?" *National Journal* (February 12, 2011): 13.

Duman, Özkan, and Dimitris Tsarouhas. "'Civilianization' in Greece versus 'Demilitarization' in Turkey: A Comparative Study of Civil-Military Relations and the Impact of the European Union." *Armed Forces & Society* 32, no. 3 (2006): 405–23.

Durkheim, Emile. *The Division of Labor in Society*. New York, NY: Free Press, 1964.

Economist. "Putin's New Model Army: Russia's Military Modernisation" (May 24, 2014).

Elgindy, Khaled. "Egypt's Troubled Transition: Elections without Democracy." *Washington Quarterly* 35, no. 2 (Spring 2012): 89–104.

Elias, Norbert. *The Civilizing Process: Sociogenetic and Psychogenetic Investigations*, trans. Edmund Jepbcott, eds. Eric Dunning, Johan Goudsblom, and Stephen Mennell. Oxford: Blackwell Publishers, 2000.

Emirates Center for Strategic Studies and Research (ECSSR). "Dr. Jamal Sanad Al-Suwaidi: Establishment of National Defense College Huge Step Forward for Military Studies and National Security" (n.d.). http://www.ecssr.ac.ae/ECSSR/appmanager/portal/ecssr?_nfls=false&_nfpb=true&lang=en&_pageLabel=basicPage&_event=showSC&scId=/StaticContents/AboutECSSR/Miscellaneous/StaticContent_0532.xml (accessed November 13, 2016).

Erkuş, Sevil. "Turkish Military Academy to Resume Student Admissions in February 2017." *Hurriyet Daily News* (November 2, 2016). http://www.hurriyetdailynews.com/Default.aspx?pageID=238&nID=105674&NewsCatID=338 (accessed November 15, 2016).

Fahmy, Khaled. "The Nation and Its Deserters: Conscription in Mehmed Ali's Egypt." In *Arming the State: Military Conscription in the Middle East and Central Asia, 1775–1925*, ed. Erik J. Zürcher, 59–78. London: I. B. Tauris, 1999.

Farrell, Theo. "Transnational Norms and Military Development: Constructing Ireland's Professional Army." *European Journal of International Relations* 7, no. 1 (2001): 37–61.

Feaver, Peter D. "The Civil-Military Problematique: Huntington, Janowitz, and the Question of Civilian Control." *Armed Forces & Society* 23, no. 2 (1996): 149–78.

———. *Armed Servants: Agency, Oversight, and Civil-Military Relations*. Cambridge, MA: Harvard University Press, 2009.

Federal Competitiveness and Statistics Authority (FCSA). "Home." http://www.fcsa.gov.ae/ (accessed November 14, 2016).

Finer, Samuel E. *The Man on Horseback: The Role of the Military in Politics*. New York, NY: Praeger, 1962.

FirstSearch. "WorldCat Advanced Search." http://newfirstsearch.oclc.org (accessed June 15, 2005).

Fitch, J. Samuel. "Military Professionalism, National Security and Democracy: Lessons from the Latin American Experience." *Pacific Focus* 4, no. 2 (1989): 99–147.

Fortescue, William. *The Third Republic in France, 1870–1940: Conflicts and Continuities*. Oxford: Psychology Press, 2000.

Gagliardo, John G. *Germany under the Old Regime, 1600–1790*. London: Longman, 1991.

Galula, David. *Counterinsurgency Warfare: Theory and Practice*. New York, NY: Greenwood Publishing Group, 2006.

Gartzke, Erik. "The Capitalist Peace." *American Journal of Political Science* 51, no. 1 (2007): 166–91.

Gat, Azar. *A History of Military Thought: From the Enlightenment to the Cold War*. Oxford: Oxford University Press, 2001.

Ghosn, Faten, Glenn Palmer, and Stuart A. Bremer. "The MID3 Data Set, 1993–2001: Procedures, Coding Rules, and Description." *Conflict Management and Peace Science* 21, no. 2 (2004): 133–54.

Gibler, Douglas M., and Meredith Reid Sarkees. "Measuring Alliances: The Correlates of War Formal Interstate Alliance Dataset, 1816–2000." *Journal of Peace Research* 41, no. 2 (2004): 211–22.

GlobalSecurity.org. "REFORGER." http://www.globalsecurity.org/military/ops/reforger.htm (accessed April 12, 2015).

Goldman, Emily O. "The Spread of Western Military Models to Ottoman Turkey and Meiji Japan." In *Sources of Military Change: Culture, Politics, Technology*, eds. Theo Farrell and Terry Terrif, 41–68. Boulder, CO: Lynne Rienner Publishers, 2002.

Görlitz, Walter, ed. *History of the German General Staff*. Boulder, CO: Westview Press, 1985.

Gorski, Philip S. "The Protestant Ethic Revisited: Disciplinary Revolution and State Formation in Holland and Prussia." *American Journal of Sociology* 99, no. 2 (1993): 265–316.

Grauer, Ryan, and Michael C. Horowitz. "What Determines Military Victory? Testing the Modern System." *Security Studies* 21, no. 1 (2012): 83–112.

Hacker, Barton C. "Engineering a New Order: Military Institutions, Technical Education, and the Rise of the Industrial State." *Technology and Culture* 34, no. 1 (1993): 1–27.

Hackett, Sir John W. *The Profession of Arms*. London: Visnews Productions, 1962.

Hale, William. *Turkish Politics and the Military*. London: Routledge, 1994.

Harries-Jenkins, Gwyn. "The Concept of Military Professionalism." Paper presented at IUS Biennial Conference, Baltimore, MD (October 1989).

Hazen, William B. *The School and the Army in Germany and France, with a Diary of Siege Life at Versailles*. New York, NY: Harper & Brothers, 1872.

Heard-Bey, Frauke. *From Trucial States to United Arab Emirates*. Dubai, UAE: Motivate Publishing, 2005.

Heiduk, Felix. "From Guardians to Democrats? Attempts to Explain Change and Continuity in the Civil-Military Relations of Post-Authoritarian Indonesia, Thailand and the Philippines." *Pacific Review* 24, no. 2 (2011): 249–71.

Henderson, William Otto. *The State and the Industrial Revolution in Prussia: 1740–1870*. Liverpool: Liverpool University Press, 1958.

———. *Studies in the Economic Policy of Frederick the Great*. Oxford: Psychology Press, 1963.

Hensel, Paul R. "ICOW Colonial History Data Set, version 1.0" (2014). http://www.paulhensel.org/icowcol.html (accessed September 15, 2016).

Heper, Metin, and Ahmet Evin, eds. *State, Democracy, and the Military: Turkey in the 1980s*. Berlin/Boston: De Gruyter, 1988.

Herrera, Geoffrey L. "Inventing the Railroad and Rifle Revolution: Information, Military Innovation and the Rise of Germany." *Journal of Strategic Studies* 27, no. 2 (2004): 243–71.

Herspring, Dale. R. *The Pentagon and the Presidency: Civil-Military Relations from FDR to George W. Bush*. Lawrence, KS: University Press of Kansas, 2005.

Hewitt, Joseph J. "A Crisis-Density Formulation for Identifying Rivalries." *Journal of Peace Research* 42, no. 2 (March 2005): 183–200.

Hintze, Otto. "Military Organization and the Organization of the State." In *The Historical Essays of Otto Hintze*, ed. F. Gilbert, 178–215. New York, NY: Oxford University Press, 1975.

Holmes, Edward Richard. *The Road to Sedan: The French Army, 1866–70*. London: Royal Historical Society, 1984.

Horne, Alistair. *The French Army and Politics, 1870–1970*. London: Macmillan Press, 1984.

Horowitz, Michael C. *The Diffusion of Military Power: The Causes and Consequences for International Politics*. Princeton, NJ: Princeton University Press, 2010.

Horowitz, Michael C., and Matthew S. Levendusky. "Drafting Support for War: Conscription and Mass Support for Warfare." *Journal of Politics* 73, no. 2 (2011): 524–34.

Huntington, Samuel P. *The Soldier and the State: The Theory and Politics of Civil-Military Relations*. Cambridge, MA: Harvard University Press, 1957.

———. *Political Order in Changing Societies*. New Haven, CT: Yale University Press, 1968.

Hurewitz, Jacob C. *Middle East Politics: The Military Dimension*. New York, NY: Frederick A. Praeger, Publishers, 1969.

Ibish, Hussein. "The UAE's Evolving National Security Strategy." Arab Gulf States Institute in Washington, Issue Paper no. 4 (April 6, 2017). http://www.agsiw.org/

wp-content/uploads/2017/04/UAE-Security_ONLINE.pdf (accessed October 3, 2017).

Inglehart, Ronald. *Modernization and Postmodernization: Cultural, Economic, and Political Change in 43 Societies*. Princeton, NJ: Princeton University Press, 1997.

International Institute for Strategic Studies (IISS). "Egypt." *Military Balance* 112, no. 1 (March 2012): 319–23.

Issawi, Charles, ed. *Economic History of the Middle East, 1800–1914: A Book of Readings*. Chicago, IL: University of Chicago Press, 1966.

Issawi, Charles. "Economic Change and Urbanization in the Middle East." In *Middle Eastern Cities: A Symposium on Ancient, Islamic, and Contemporary Middle Eastern Urbanism*, ed. Ira M. Lapidus, 102–120. Berkeley, CA: University of California Press, 1979.

———. *Economic History of Turkey: 1800–1914*. Chicago: University of Chicago Press, 1980.

———. *An Economic History of the Middle East and North Africa*. New York, NY: Columbia University Press, 1982.

Janowitz, Morris. *The Professional Soldier: A Social and Political Portrait*. New York, NY: Free Press, 1971.

———. "Some Observations on the Comparative Analysis of Middle Eastern Military Institutions." In *War, Technology and Society in the Middle East*, eds. V. J. Parry and M. E. Yapp, 412–40. London: Oxford University Press, 1975.

Jenkins, Gareth. "Continuity and Change: Prospects for Civil-Military Relations in Turkey." *International Affairs* 83 (2007): 339–55.

Johnson, David E., Jennifer D. P. Moroney, and Roger Cliff. *Preparing and Training for the Full Spectrum of Military Challenges: Insights from the Experiences of China, France, the United Kingdom, India, and Israel*. Santa Monica, CA: RAND Corporation, 2009.

Johnston, Alastair Iain. *Cultural Realism: Strategic Culture and Grand Strategy in Chinese History*. Princeton, NJ: Princeton University Press, 1998.

Joint Chiefs of Staff. *National Military Strategic Plan for the War on Terrorism*. Washington, DC: Government Printing Office, 2006.

Joint Command and Staff College (JCSC). "عن الكلية" ("About the College"). http://www.jcsc.milnet.ae (accessed April 5, 2016).

Karpat, Kemal H. "Military Interventions: Army-Civilian Relations in Turkey Before and After 1980." In *State, Democracy, and the Military: Turkey in the 1980s*, eds. Metin Heper and Ahmet Evin, 137–58. Berlin/Boston: De Gruyter, 1988.

Karsh, Efraim, and Inari Karsh. *Empires of the Sand: The Struggle for Mastery in the Middle East, 1789–1923*. Cambridge, MA: Harvard University Press, 1999.

Katzenstein, Peter J. *Cultural Norms and National Security: Police and Military in Postwar Japan*. Ithaca, NY: Cornell University Press, 1996.

Keegan, John, ed. *World Armies*, 2nd ed. Detroit, MI: Gale Research Company, 1983.

———. *A History of Warfare*. New York, NY: Alfred A. Knopf, 1993.

Kelley, Kevin P., and Joan Johnson-Freese. "Rethinking Professional Military Education." *E-Notes*, Foreign Policy Research Institute (October 25, 2013).

http://www.fpri.org/articles/2013/10/rethinking-professional-military-education (accessed March 18, 2015).

Kentor, Jeffrey, and Edward Kick. "Bringing the Military Back in: Military Expenditures and Economic Growth, 1990 to 2003." *Journal of World-Systems Research* 14, no. 2 (2015): 142–72.

Khaleej Times. "Shaikha Shamsa Calls for Military Education in Schools in UAE" (June 4, 2011). http://www.khaleejtimes.com/article/20110603/ARTICLE/306039989/1002 (accessed August 8, 2016).

Kier, Elizabeth. *Imagining War: French and British Military Doctrine between the Wars*. Princeton, NJ: Princeton University Press, 1999.

Kilcullen, David. *The Accidental Guerrilla: Fighting Small Wars in the Midst of a Big One*. Oxford: Oxford University Press, 2009.

King, Gary, Michael Tomz, and Jason Wittenberg. "Making the Most of Statistical Analyses: Improving Interpretation and Presentation." *American Journal of Political Science* 44, no. 2 (April 2000): 347–61.

Kiser, Edgar, and Joachim Schneider. "Bureaucracy and Efficiency: An Analysis of Taxation in Early Modern Prussia." *American Sociological Review* 59, no. 2 (1994): 187–204.

Klein, James P., Gary Goertz, and Paul F. Diehl. "The New Rivalry Dataset: Procedures and Patterns." *Journal of Peace Research* 43, no. 3 (May 2006): 331–48.

Kollias, Christos G. "The Greek-Turkish Conflict and Greek Military Expenditure, 1960–92." *Journal of Peace Research* 33, no. 2 (1996): 217–28.

Kovacs, Arpad F. "French Military Institutions before the Franco-Prussian War." *American Historical Review* 51, no. 2 (1946): 217–35.

Kuczynski, Jürgen. *Germany 1800 to the Present Day: A Short History of Labour Conditions*. London: Frederick Muller, Limited, 1945.

Labbé, Morgane. "Institutionalizing the Statistics of Nationality in Prussia in the 19th Century (from Local Bureaucracy to State-level Census of Population)." *Centaurus* 49, no. 4 (2007): 289–306.

Lahlou, Alia. "SCAF Member Defends Transition Rule in Egypt." *Washington Report on Middle East Affairs* 30, no. 8 (2011): 68.

Lamb, Christopher J., and Brittany Porro. "Next Steps for Transforming Education at National Defense University." *Joint Force Quarterly* 76 (1st Quarter 2015): 40–47.

Landes, David. *The Wealth and Poverty of Nations: Why Some Are So Rich and Some So Poor*. New York, NY: W. W. Norton and Company, 1999.

Lebovics, Herman. *The Alliance of Iron and Wheat in the Third French Republic, 1860–1914*. Baton Rouge, LA: Louisiana State University Press, 1988.

Lee, Dongmin. "Chinese Civil-Military Relations." *Armed Forces & Society* 32, no. 3 (2006): 437–53.

Lerner, Daniel, and Lucille W. Pevsner. *The Passing of Traditional Society: Modernizing the Middle East*. New York, NY: Free Press of Glencoe, 1964.

Lewis, Bernard. *The Emergence of Modern Turkey*, 2nd ed. London: Oxford University Press, 1968.

———. *What Went Wrong? Western Impact and Middle Eastern Response*. London: Oxford University Press, 2002.

Libel, Tamir. "From the People's Army to the Jewish People's Army: The IDF's Force Structure between Professionalization and Militarization." *Defense & Security Analysis* 29, no. 4 (2013): 280–92.

Lohmann, Ingrid, and Christine Mayer. "Educating the Citizen: Two Case Studies on Inclusion and Exclusion in Prussia in the Early Nineteenth Century." *Paedagogica Historica* 43, no. 1 (2007): 7–27.

Losey, Stephen. "Red Flag Training Exercise to Begin at Nellis Monday." *Air Force Times* (January 24, 2015). http://www.airforcetimes.com/story/military/tech/2015/01/24/red-flag-training-exercise-to-begin-at-nellis-monday/22222799/ (accessed April 14, 2015).

Lutterbeck, Derek. "Arab Uprisings, Armed Forces, and Civil–Military Relations." *Armed Forces & Society* 39, no. 1 (2013): 28–52.

Lynch, Marc. "America and Egypt After the Uprisings." *Survival* 53, no. 2 (March 2011): 31–42.

———. *The Arab Uprising: The Unfinished Revolutions of the New Middle East*. New York, NY: PublicAffairs, 2012.

Lynn, John Albert. *The Bayonets of the Republic: Motivation and Tactics in the Army of Revolutionary France, 1791–94*. Boulder, CO: Westview Press, 1996.

———. *Battle: A History of Combat and Culture*. Boulder, CO: Westview Press, 2003.

Lyons, Gene M. "The New Civil-Military Relations." *American Political Science Review* 55, no. 1 (1961): 53–63.

Maddison Project. "Database," 2013 version. http://www.ggdc.net/maddison/maddison-project/home.htm (accessed August 15, 2016).

Maniruzzaman, Talukder. *Military Withdrawal from Politics: A Comparative Study*. Cambridge, MA: Ballinger, 1987.

Matheny, Michael R. *Carrying the War to the Enemy: American Operational Art to 1945*. Norman, OK: University of Oklahoma Press, 2011.

Mattair, Thomas R. *The Three Occupied UAE Islands: The Tunbs and Abu Musa*. Abu Dhabi, UAE: Emirates Center for Strategic Studies and Research, 2005.

McGhee, George. *The US-Turkish–NATO Middle East Connection: How the Truman Doctrine Contained the Soviets in the Middle East*. New York, NY: St. Martin's Press, 1990.

McGivern, R. Mark. "What the Hell Is Happening at NDU? (III)." *Best Defense* (September 8, 2014). http://foreignpolicy.com/2014/09/08/what-the-hell-is-happening-at-ndu-iii/ (accessed March 18, 2015).

McGrath, Cam. "The Generals Guard Their Economic Empire." *Middle East* 434 (2012): 30–31.

McGregor, Andrew. "Egypt, the UAE, and Arab Military Intervention in Libya." *Terrorism Monitor* 12, no. 17 (2014). https://jamestown.org/program/egypt-the-uae-and-arab-military-intervention-in-libya/ (accessed October 9, 2016).

McNeill, Sophie. "Retired Australian Major General Mike Hindmarsh Faces Questions about Knowledge of Civilian Attacks in Yemen." *Australian Broadcasting Corporation* (February 15, 2016). http://www.abc.net.au/news/2016-02-08/

yemen-civil-war-questions-role-australian-soldier-mike-hindmarsh/7141638 (accessed November 14, 2016).

Metz, Steven. "Strategic Horizons: U.S. Professional Military Education on the Chopping Block." *World Politics Review* (April 17, 2013). http://www.ndu.edu/Portals/59/Documents/BOV_Documents/2013/Metz--US%20PME%20on%20the%20Chopping%20Block.pdf (accessed March 18, 2015).

Middle East Channel. "The State of the Egyptian Revolution." *Project on Middle East Political Science* (September 21, 2011). http://www.pomeps.org/wp-content/uploads/2011/09/POMEPS_BriefBooklet6_Egypt_web.pdf (accessed August 21, 2012).

Millett, Allan R., Williamson Murray, and Kenneth H. Watman. "The Effectiveness of Military Organizations." *International Security* 11, no. 1 (1986): 37–71.

Mitchell, Allan. *Victors and Vanquished: The German Influence on Army and Church in France after 1870*, vol. 2. Chapel Hill, NC: University of North Carolina Press, 1984.

Moroney, Jennifer D. P., Nancy E. Blacker, Renee Buhr, James McFadden, Cathryn Quantic Thurston, and Anny Wong. *Building Partner Capabilities for Coalition Operations*. Santa Monica, CA: RAND Arroyo Center, 2007.

Morrow, James D. "Arms versus Allies: Trade-Offs in the Search for Security." *International Organization* 47, no. 2 (1993): 207–33.

Nagl, John. *Learning to Eat Soup with a Knife: Counterinsurgency Lessons from Malaya and Vietnam*. Chicago, IL: University of Chicago Press, 2005.

Naidu, Suresh, Yaw Nyarko, and Shing-Yi Wang. "Monopsony Power in Migrant Labor Markets: Evidence from the United Arab Emirates." Unpublished manuscript (2015). http://assets.wharton.upenn.edu/~was/UAElabormobility.pdf (accessed November 14, 2016).

Nassif, Hicham Bou. "Wedded to Mubarak: The Second Careers and Financial Rewards of Egypt's Military Elite, 1981–2011." *Middle East Journal* 67, no. 4 (2013): 509–30.

National Defense College. "Delegation from Near and South Asia Centre for Strategic Studies (NESA) Visits National Defense College" (2013). http://www.ndc.ac.ae/en/delegation-from-near-and-south-asia-centre-for-strategic-studies-nesa-visits-national-defense-college (accessed November 14, 2016).

———. "Welcome to the UAE NDC." http://www.ndc.ac.ae/en/ (accessed November 13, 2016).

National Training Center (NTC). "National Training Center, Fort Irwin" (2015). http://www.irwin.army.mil/Pages/Units/NTC/NTC.html (accessed April 14, 2015).

Nickerson, Hoffman. *The Armed Horde, 1793–1939*. New York, NY: G. P. Putnam's Sons, 1940.

Nordlinger, Eric A. *Soldiers in Politics: Military Coups and Governments*. Englewood Cliffs, NJ: Prentice-Hall, 1977.

Novichev, A. D. "Ocherki ekonomiki Turtsii." In *The Economic History of the Middle East, 1800–1914: A Book of Readings*, ed. Charles Issawi, 65–70. Chicago, IL: University of Chicago Press, 1966.

Nunn, Frederick M. *Yesterday's Soldiers: European Military Professionalism in South America, 1890–1940*. Lincoln, NE: University of Nebraska Press, 1983.

———. *The Time of the Generals: Latin American Professional Militarism in World Perspective*. Lincoln, NE: University of Nebraska Press, 1992.

O'Donnell, Guillermo, Philippe C. Schmitter, and Laurence Whitehead, eds. *Transitions from Authoritarian Rule*. Baltimore, MD: Johns Hopkins University Press, 1986.

Oktay, Sibel. "Decade of Ebb and Flow: Civil-Military Relations in Turkey between 1997 and 2007." Paper presented at the Annual Meeting of the International Studies Association (2010).

Osman, Julia. "Ancient Warriors on Modern Soil: French Military Reform and American Military Images in Eighteenth-Century France." *French History* 22, no. 2 (2008): 175–96.

Owen, Roger. *The Middle East in the World Economy, 1800–1914*. London: Methuen, 1981.

Palmer, Glenn, Vito D'Orazio, Michael Kenwick, and Matthew Lane. "The MID4 Data Set: Procedures, Coding Rules, and Description." *Conflict Management and Peace Science* 32, no. 2 (2015): 222–42.

Paparone, Christopher R., Ruth A. Anderson, and Reuben R. McDaniel, Jr. "Where Military Professionalism Meets Complexity Science." *Armed Forces & Society* 34, no. 3 (2008): 433–49.

Pascal, Anthony H., Michael Kennedy, and Steven P. Rosen. *Men and Arms in the Middle East: The Human Factor in Military Modernization*. Report no. R-2460-NA. Santa Monica, CA: RAND Corporation, 1979.

Pascal, Anthony H. *Are Third World Armies Third Rate? Human Capital and Organizational Impediments to Military Effectiveness*. Report no. P-6433. Santa Monica, CA: RAND Corporation, 1980.

Perlmutter, Amos, and Valerie Plave Bennett. *The Political Influence of the Military: A Comparative Reader*. New Haven, CT: Yale University Press, 1980.

Peters, Ralph. "Learning to Lose." *American Interest* 2, no. 6 (July 2007). http://www.the-american-interest.com/2007/07/01/learning-to-lose/ (accessed March 18, 2015).

Pichichero, Christy. "Le Soldat Sensible: Military Psychology and Social Egalitarianism in the Enlightenment French Army." *French Historical Studies* 31, no. 4 (2008): 553–80.

Pilster, Ulrich, and Tobias Böhmelt. "Coup-Proofing and Military Effectiveness in Interstate Wars, 1967–99." *Conflict Management and Peace Science* 28, no. 4 (2011): 331–50.

Pitman, Paul M., III. *Turkey: A Country Study*. Library of Congress, Federal Research Division, Area Handbook Series. Washington, DC: Library of Congress, 1988.

Pollack, Kenneth M. "The Influence of Arab Culture on Arab Military Effectiveness." Ph.D. Diss., Massachusetts Institute of Technology, 1996.

———. *Arabs at War: Military Effectiveness, 1948–1991*. Lincoln, NE: University of Nebraska Press, 2002.

Porch, Douglas. *The March to the Marne: The French Army, 1871–1914*. Cambridge: Cambridge University Press, 1981.

Posen, Barry R. "Nationalism, the Mass Army, and Military Power." *International Security* 18, no. 2 (1993): 80–124.

Powell, Jonathan. "Determinants of the Attempting and Outcome of Coups d'Etat." *Journal of Conflict Resolution* 56, no. 6 (2012): 1017–40.

Preston, Richard A. *Perspectives in the History of Military Education and Professionalism*. Technical Report 22. Colorado Springs, CO: United States Air Force Academy, 1980.

Price, Roger. *An Economic History of Modern France, 1730–1914*. New York, NY: St. Martin's Press, 1981.

Ralston, David B. *Importing the European Army: The Introduction of European Military Techniques and Institutions into the Extra-European World, 1600–1914*. Chicago, IL: University of Chicago Press, 1990.

Ramsay, Kristopher W. "Settling It on the Field: Battlefield Events and War Termination." *Journal of Conflict Resolution* 52, no. 6 (2008): 850–79.

Reiter, Daniel, and Allan C. Stam, III. "Democracy and Battlefield Military Effectiveness." *Journal of Conflict Resolution* 42, no. 3 (1998): 259–77.

Renshon, Jonathan, and Arthur Spirling. "Modeling 'Effectiveness' in International Relations." *Journal of Conflict Resolution* 59, no. 2 (2015): 207–38.

Resende-Santos, João. "Anarchy and the Emulation of Military Systems: Military Organization and Technology in South America, 1870–1930." *Security Studies* 5, no. 3 (1996): 193–260.

Richards, Alan, and John Waterbury. *A Political Economy of the Middle East: State, Class, and Economic Development*. Boulder, CO: Westview Press, 1990.

Riches, Daniel. "Early Modern Military Reform and the Connection Between Sweden and Brandenburg-Prussia." *Scandinavian Studies* 77, no. 3 (2005): 347–64.

Ricks, Thomas. *The Generals: American Military Command from World War II to Today*. New York, NY: Penguin, 2013.

Rosen, Stephen P. "Military Effectiveness: Why Society Matters." *International Security* 19, no. 4 (1995): 5–31.

———. *Societies and Military Power: India and Its Armies*. Ithaca, NY: Cornell University Press, 1996.

Rosinski, Herbert. *The German Army*. New York, NY: Praeger, 1939.

Rowe, Michael. "France, Prussia, or Germany? The Napoleonic Wars and Shifting Allegiances in the Rhineland." *Central European History* 39, no. 4 (2006): 611–40.

Sadık, Giray. "Comparative Analysis of Democratization and Civil-Military Relations in Turkey and South Korea Till Early 2000s." *Journal of Gazi Academic View* 6, no. 12 (2013): 205–22.

Sagan, Eli. *Citizens & Cannibals: The French Revolution, the Struggle for Modernity, and the Origins of Ideological Terror*. Lanham, MD: Rowman & Littlefield, 2001.

Salem, Ola. "New Defence College to Train UAE's Future Leaders." *National* (December 31, 2012). http://www.thenational.ae/news/uae-news/new-defence-college-to-train-uaes-future-leaders (accessed November 13, 2016).

Sarç, O. C. "Tanzimat ve Sanayimiz." In *The Economic History of the Middle East, 1800–1914: A Book of Readings*, ed. Charles Issawi, 46–59. Chicago, IL: University of Chicago Press, 1966.

Sarkees, Meredith Reid, and Frank Wayman. *Resort to War: 1816–2007*. Washington, DC: CQ Press, 2010.

Sarkesian, Sam C. *Beyond the Battlefield: The New Military Professionalism*. New York, NY: Pergamon Press, 1981.

———. "Professionalism, Military." In *International Military and Defense Encyclopedia*, ed. Trevor N. Dupuy, 2193–97. Washington, DC: Brassey's, 1993.

Sayigh, Yezid. *Above the State: The Officers' Republic in Egypt*. Washington, DC: Carnegie Endowment for International Peace (August 2012). http://carnegieendowment.org/files/officers_republic.pdf (accessed August 19, 2015).

Schifferle, Peter. *America's School for War: Fort Leavenworth, Officer Education, and Victory in World War II*. Lawrence, KS: University Press of Kansas, 2010.

Schleunes, Karl A. "Enlightenment, Reform, Reaction: The Schooling Revolution in Prussia." *Central European History* 12, no. 4 (1979): 315–42.

Shanahan, William Oswald. *Prussian Military Reforms 1786–1813*. Issue 520 of Columbia Studies in the Social Sciences. Norwalk, CT: AMS Press, 1966.

Shaw, Stanford. *Between Old and New: The Ottoman Empire under Sultan Selim III, 1789–1807*. Cambridge, MA: Harvard University Press, 1971.

———. *Empire of the Gazis: The Rise and Decline of the Ottoman Empire, 1280–1808, vol. 1 of History of the Ottoman Empire and Modern Turkey*. Cambridge: Cambridge University Press, 1976.

Showalter, Dennis E. "Diplomacy and the Military in France and Prussia, 1870." *Central European History* 4 (1971): 346–53.

Simon, Sheldon W., ed. *The Military and Security in the Third World: Domestic and International Impacts*. Boulder, CO: Westview Press, 1978.

Singer, J. David, Stuart Bremer, and John Stuckey. "Capability Distribution, Uncertainty, and Major Power War, 1820–1965." In *Peace, War, and Numbers*, ed. Bruce Russett, 19–48. Beverly Hills, CA: Sage, 1972.

Singer, J. David. "Reconstructing the Correlates of War Dataset on Material Capabilities of States, 1816–1985." *International Interactions* 14, no. 2 (1987): 115–32.

Smith, Adam. *An Inquiry into the Nature and Causes of the Wealth of Nations: A Selected Edition*. Oxford: Oxford University Press, 1998 [1776].

Smith, Mark. *NATO Enlargement during the Cold War: Strategy and System in the Western Alliance*. New York, NY: Palgrave, 2000.

Sørensen, Henning. "New Perspectives on the Military Profession: The I/O Model and Esprit de Corps Reevaluated." *Armed Forces & Society* 20, no. 4 (1994): 599–617.

Springborg, Robert. "Economic Involvements of Militaries." *International Journal of Middle East Studies* 43, no. 3 (2011a): 397–99.

———. "The Precarious Economics of Arab Springs." *Survival* 53, no. 6 (2011b): 85–104.

Stein, Aaron. "This Is How Turkey's Incursion into Syria Could Get Bogged Down." *War on the Rocks* (September 7, 2016). http://warontherocks.com/2016/09/

this-is-how-turkeys-incursion-into-syria-could-get-bogged-down/ (accessed November 15, 2016).

Stepan, Alfred. *The Military in Politics: Changing Patterns in Brazil*. Princeton, NJ: Princeton University Press, 1971.

———. *Rethinking Military Politics: Brazil and the Southern Cone*. Princeton, NJ: Princeton University Press, 1988.

Talmadge, Caitlin. *The Dictator's Army: Battlefield Effectiveness in Authoritarian Regimes*. Ithaca, NY: Cornell University Press, 2015.

Teitler, Gerke. *The Genesis of the Professional Officers Corps*. Beverly Hills, CA: Sage, 1977.

Tilly, Charles. *Coercion, Capital, and European States, A.D. 990–1992*. Cambridge, MA: Blackwell, 1992.

Tombs, Robert. *France, 1814–1914*. London: Longman, 1996.

Tomz, Michael, Jason Wittenberg, and Gary King. "CLARIFY: Software for Interpreting and Presenting Statistical Results," version 2.1. Stanford University, University of Wisconsin, and Harvard University (January 5, 2003). http://gking.harvard.edu (accessed September 15, 2016).

Toronto, Nathan W. "Training Is Not Enough: Economic Development and Military Professionalism in Turkey." *St. Antony's International Review* 7, no. 1 (2011a): 119–43.

———. "Active Inaction: Interagency Security Assistance to Egypt." *Interagency Paper*, no. 6, Col. Arthur D. Simons Center for the Study of Interagency Cooperation (December 2011b).

———. "Compulsory Military Service: Some Lessons from History." *National Defense* 1 (June 2014): 43–47.

———. "Why Professionalize? Economic Modernization, Security, and Military Professionalism." *Foreign Policy Analysis* 13, no. 4 (October 2017): 854–75.

Trinquier, Roger. *Modern Warfare: A French View of Counterinsurgency*. New York, NY: Greenwood Publishing Group, 2006 [1961].

Ubicini, M. A. "Letters on Turkey." In *The Economic History of the Middle East, 1800–1914: A Book of Readings*, ed. Charles Issawi, 41–45. Chicago, IL: University of Chicago Press, 1966.

Ulrich's Periodicals Directory. http://www.ulrichsweb.com/ulrichsweb/ (accessed June 15, 2005).

UK Ministry of Defence (MoD). *Securing Britain in an Age of Uncertainty: The Strategic Defence and Security Review*. London: UK Ministry of Defence, October 2010.

Upton, Emory. *Armies of Asia and Europe: Embracing Official Reports: Armies of Japan, China, India, Persia, Italy, Russia, Austria, Germany, France, and England*. New York, NY: D. Appleton and Company, 1878.

United Arab Emirates (UAE). "The President, His Highness Shaikh Khalifa bin Zayed bin Sultan Al Nahyan, Has Issued a Decree Establishing a National Defense College" (August 1, 2012). http://www.ndc.ac.ae/en/inuguration (accessed November 13, 2016).

———. “Vision 2021.” https://www.vision2021.ae/en (accessed November 14, 2016).

United Nations Development Programme (UNDP). “Arab Human Development Reports.” http://www.arab-hdr.org/ (accessed November 14, 2016).

US Department of Defense (DoD). *National Defense Strategy of the United States of America*. Washington, DC: Government Printing Office, 2005.

———. *Quadrennial Defense Review Report*. Washington, DC: Government Printing Office, 2006.

———. *Sustaining US Global Leadership: Priorities for 21st Century Defense*. Washington, DC: US Department of Defense, January 5, 2012.

Vagts, Alfred. “The Military and Politics.” In *War*, ed. Lawrence Freedman, 127–30. New York, NY: Oxford University Press, 1994.

Waltz, Kenneth N. *Theory of International Politics*. Reading, MA: Addison-Wesley Publishing Company, 1979.

Watson, Cynthia A. *Military Education: A Reference Handbook*. New York, NY: Greenwood Publishing Group, 2007.

Weber, Max. “The Bureaucratization of Politics and the Economy.” In *Essays in Economic Sociology*, ed. Richard Swedberg, 109–15. Princeton: Princeton University Press, 1999a.

———. “The Three Types of Legitimate Domination.” In *Essays in Economic Sociology*, ed. Richard Swedberg, 99–108. Princeton: Princeton University Press, 1999b.

Welch, Claude E., Jr. *No Farewell to Arms: Military Disengagement from Politics in Africa and Latin America*. Boulder, CO: Westview Press, 1987.

———. “Military Disengagement from Politics: Paradigms, Processes, or Random Events.” *Armed Forces & Society* 18, no. 3 (1992): 323–42.

White House. *National Security Strategy*. Washington, DC: Government Printing Office, May 2010.

World Bank. “Education Statistics (EdStats) Database.” http://datatopics.worldbank.org/education/ (accessed September 15, 2016).

Yildirim, A. Kadir. “Military, Political Islam, and the Future of Democracy in Egypt.” *Insight Turkey* 15, no. 4 (2013): 61–68.

Yinanç, Barçın. “Post-Coup Attempt Restructuring Means Turkish Army May No Longer Be Functional: Özcan.” *Hurriyet Daily News* (November 7, 2016). http://www.hurriyetdailynews.com/Default.aspx?pageID=238&nID=105805&NewsCatID=338 (accessed November 15, 2016).

Index

Page references for figures and tables are italicized.

About the Author

Nathan W. Toronto has taught strategy and security studies to military officers and government officials for over ten years, at the US Army School of Advanced Military Studies and the United Arab Emirates National Defense College. Dr. Toronto's research interests include Middle East security, theories of warfare, and civil-military relations, and his work has appeared in such outlets as *Journal of Conflict Resolution*, *Armed Forces & Society*, and *Foreign Policy Analysis*. He also writes science fiction, and can be found online at www.nathantoronto.com and @NathanToronto. The views expressed in this work are his own and do not necessarily represent the views of the UAE National Defense College or the UAE Government.